LJ4

GOOD HOUSEKEEPING
PARTIES
FROM YOUR
FREEZER

GOOD HOUSEKEEPING
PARTIES FROM YOUR FREEZER

by

GOOD HOUSEKEEPING
INSTITUTE

illustrated by
VANESSA LUFF

Ebury Press · London

Published by Ebury Press
Chestergate House
Vauxhall Bridge Road
London SW1V 1HF

First impression 1977

© The National Magazine Co Ltd 1977

ISBN 0 85223 112 1

Editor Amanda Atha
Designer Derek Morrison
Cover photograph by Roger Phillips

*GH Parties from your Freezer is based
on GH Freezer recipes*

Filmset and printed in Great Britain by
BAS Printers Limited, Wallop, Hampshire
and bound by
Mansell (Bookbinders) Ltd
Witham, Essex

Contents

Introduction

Freezers make life easier for most people but for party givers they are of special value. You can prepare, cook and freeze your party menu days ahead. All that is left to do on the day is to get the food out of the freezer, collect the fresh salads and vegetables and last minute garnishes for the frozen dishes. No standing over the stove all day and appearing exhausted to greet your guests.

We have planned menus ranging from dinners for four to buffets for forty and given detailed notes on when to take the food from the freezer, when to put it in the oven and how to garnish it.

The first section of the book deals with dinner party menus for four to eight people. These seem to be the most popular numbers for entertaining. Each menu has been worked out carefully to cater for the number given in the menu but you may want to vary the number of servings. Where we have suggested freezing the food in individual packs or the main course is of separate portions like Chicken Kiev it is not difficult to adjust the quantities. With casseroles we have suggested freezing in two portions (four and four) in some of the menus so again there is no difficulty. If you read the menus carefully you will find quite a lot of scope for variations in number. The important thing to remember is to label your wrapped food clearly with the number of portions before you freeze it. For the larger menus we tell you how many people each dish will serve so it is not difficult to chop and change. Thawing times are important – it's something you can't hurry. Slow thawing in the refrigerator is far better than emergency methods like putting the food in warm water so if you can't trust yourself to get up early enough in the morning, take the food out of the freezer the night before and put it in the refrigerator to thaw.

We hope this book will give you many successful parties but if you have any queries about any of the recipes do write to us enclosing a stamped addressed envelope.

CAROL MACARTNEY
Director

Good Housekeeping Institute
Chestergate House
Vauxhall Bridge Road
London SW1V 1HF

Dinner from Your Freezer

FOR FOUR

Carrot and courgette soup
Cider beef with mushrooms
*Mange-tout   *Creamed potatoes
Black cherry sorbet

*Starred dishes are recipe suggestions or accompaniments
not suitable for freezing*

Countdown

The night before   Remove the cider beef from the freezer. Unwrap and place in a casserole. Allow to thaw overnight in the refrigerator.
Late afternoon   Remove soup from the freezer, turn into a saucepan and allow to thaw. Prepare potatoes.
Before guests arrive   Place casserole in oven and cook according to instructions. Cook potatoes and mange-tout. Heat the soup according to instructions. Remove the sorbet from the freezer, place in refrigerator and allow to soften before placing in glass dishes.

Carrot and courgette soup

2 oz butter
½ lb carrots, pared and sliced
½ lb courgettes, trimmed and sliced
1 pt well flavoured stock
1 bay leaf
1 level tbsp tomato paste
1 level tbsp caster sugar
salt and freshly ground black pepper

To make : Melt the butter in a saucepan, add the carrots and courgettes, cover and cook gently for about 10 minutes, until soft. Pour in the stock and remaining ingredients. Bring the soup to the boil and simmer for 30 minutes, remove the bay leaf then pour it into a blender and blend for a few

seconds; the soup should not be completely smooth. Chill.
To pack and freeze : Pour into a rigid container, seal, label and freeze.
To use : Thaw at room temperature for 6–8 hours. Turn the soup into a saucepan and heat gently until thawed. Bring to the boil before serving.

Cider beef with mushrooms

1½ lb shin of beef, trimmed and cubed
3 onions, skinned and sliced
½ pt cider
½ oz margarine
1 level tbsp flour
¼ lb button mushrooms
salt and pepper

To make : Marinade the meat and onions in the cider overnight. Drain and reserve the marinade. Melt the margarine in a frying pan and sauté the meat and onion until the meat is brown and sealed. Transfer to a 2½-pt casserole. Remove the pan from the heat and stir the flour into the sediment in the frying pan, then gradually stir in the cider marinade. Return to the heat, bring to the boil, add the mushrooms and season with salt and pepper. Pour over the meat, cover and cook in the oven at 325°F (mark 3) for 3 hours. Cool rapidly.
To pack and freeze : Turn the meat into a rigid foil or polythene container and cover with a lid, leaving ½ in headspace. Overwrap with polythene if

9

FOR DOUBLE QUANTITY
USE 1 x 2 4 oz. tin purée.

Black cherry sorbet

Mange-tout

This recipe is a menu suggestion not suitable for freezing

1 lb fresh or frozen mange-tout
1 oz butter
2 oz mushrooms, thickly sliced
3 tomatoes, skinned and quartered
salt and freshly ground black pepper

Cook the mange-tout in boiling salted water and drain well. Melt the butter in a frying pan and when hot sauté the mushrooms for a few minutes. Add the mange-tout and tomatoes to the pan, mix well and season with salt and pepper.

Black cherry sorbet

15-oz can black cherries, stoned and
    puréed
2 tbsps Kirsch
1 tbsp lemon juice
1 egg white

To make : Combine the cherry purée, Kirsch and lemon juice. Pour into a container and freeze until mushy. Whisk the egg white, turn the cherry mixture into a bowl, and fold the stiffly whisked egg white into the mixture. Return to the container and freeze until firm.
To pack : Seal and label the container and return it to the freezer.
To use : Allow to 'come to' in the refrigerator for about 2 hours then scoop into glass dishes.

necessary, seal, label and freeze. Alternatively line a clean casserole dish with foil, fill with the meat mixture and freeze until firm; remove the foil covered block from the casserole, fold over the foil and overwrap with freezewrap. Seal, label and return it to the freezer.
To use : Unwrap and turn the meat into a casserole, or leave in a foil container. *Either* Thaw overnight in the refrigerator and reheat in the oven at 375°F (mark 5) for about 45 minutes. *Or* Cook from frozen at 375°F (mark 5) for 2 hours.

FOR FOUR

Courgettes à la grecque

Chicken Kiev   *New minted potatoes
*Dressed green salad or broccoli

Apricot soufflé

*Starred dishes are recipe suggestions or accompaniments
not suitable for freezing*

Countdown

The night before   Remove the courgettes from the freezer, loosen the seal, set the freezer bag in a bowl in the refrigerator.

Late afternoon Take the soufflé from the freezer, unwrap and leave at room temperature. Prepare the vegetables and salad ingredients, placing salad items in a freezer bag in the refrigerator to crisp. Decorate soufflé. Peel and seed tomatoes. Chop parsley for courgettes.

Before guests arrive Pour oil into a deep pan for the Chicken Kiev. Turn the courgettes into a serving dish. Arrange the salad. Cook the potatoes. Cook the Chicken Kiev and keep warm.

When guests arrive Dress salad (see page 83, French dressing) and serve.

Courgettes à la grecque

2 small onions, skinned and thinly sliced
3 tbsps olive oil
1 clove garlic, skinned and crushed
just over $\frac{1}{4}$ pt dry white wine
seasoning
$1\frac{1}{2}$ lb courgettes
$\frac{1}{2}$ lb tomatoes
pinch of dried chervil, optional

To complete from freezer
$\frac{1}{2}$ lb tomatoes
chopped fresh chervil or parsley

To make: Sauté the onions in the hot oil until soft but not coloured; add garlic, wine and a little seasoning. Wipe the courgettes and discard the end slices. Cut the remainder into rings. Skin and quarter the tomatoes, discarding the seeds. Add the courgettes and tomatoes to the pan and cook gently, without covering, for 10 minutes. Cool quickly. Add a little chervil if liked (or add fresh chervil, if available, when serving).

To pack and freeze: Spoon into a freezer bag, set in a rigid container to give it shape. Put in freezer. When set in shape, remove from container and overwrap with foil, or place inside another freezer bag, seal and label. Return to freezer.

To thaw: Leave overnight in the refrigerator.

To serve: Add a further $\frac{1}{2}$ lb skinned and seeded tomatoes. Adjust seasoning if necessary. Sprinkle with chopped fresh chervil (if available, and provided dried chervil was not included before freezing) or with chopped parsley.

Chicken Kiev

4 fresh chicken breasts
2 oz garlic butter *(see page 81)*
1 oz seasoned flour
2 eggs, beaten
6 oz fresh white breadcrumbs

To complete from freezer
sprigs of watercress

To make: Beat out the flesh of the chicken breasts like escalopes. Make the garlic butter, form into a sausage shape and pop into the freezer to set. Divide the butter into 4 pieces, place on the chicken breasts. Roll up and secure with cocktail sticks. Coat in seasoned flour, dip in egg, then toss in breadcrumbs. Repeat the egg and crumbing process a second time.

Chicken Kiev

To pack and freeze: Place on a small baking tray inside a large freezer bag and put in the freezer. When the chicken is completely frozen, wrap up individually in heavy duty foil, place in a freezer bag, seal and label. Store for up to one month.

To thaw: Allow 8 hours in refrigerator, or cook from frozen as directed below.

To serve: When thawed deep fry at 350°F for about 10 minutes. If still frozen, allow about 20 minutes, depending on size; fry at 350°F to start, then reduce to 300°F. Remove cocktail sticks. Prick with a fork before serving to avoid butter spluttering. Garnish with sprigs of watercress.

Apricot soufflé

15-oz can apricot halves, drained
5 egg yolks
6 level tbsps caster sugar
1 level tbsp powered gelatine
2 tbsps water
5 tbsps apricot brandy
2 tbsps lemon juice
3 egg whites

To complete from freezer
¼ pt double or whipping cream

To make: Line a 6-in (1½-pt) soufflé dish with non-stick paper to extend 3 in above rim. Sieve or purée the apricots, add the apricot juice. Place with the egg yolks and sugar in a basin over a pan of hot water and whisk until the mixture thickens and leaves a trail when the whisk is removed.

Dissolve the gelatine in the water; add the apricot brandy and lemon juice. Stir a little of the whisked mixture into the gelatine, return it to the bulk of the whisked mixture and fold in. Chill until on the point of setting. Stiffly whisk the egg whites and fold into the apricot mixture. Turn into the prepared dish and allow to set.

To pack and freeze: Cover the dish loosely with foil and put in the freezer. When quite frozen, remove foil and place the dish inside a freezer bag. Seal and label. Place carefully in the freezer so that no heavy weights will rest on it.

To thaw: Allow 12 hours in a refrigerator or 4–5 hours at room temperature.

To serve: Remove the paper collar. Decorate with whipped cream.

FOR FOUR

*Fresh tomato salad with French dressing

Chicken with curry and lemon
*Boiled rice
*Cucumber slices with yoghurt dressing

Meringues Chantilly

Starred dishes are recipe suggestions or accompaniments not suitable for freezing

Countdown

The day before   Take chicken with curry and lemon from the freezer if it is to be thawed before cooking – the best method. It can, however, be cooked from frozen – see recipe.

During the afternoon   Take the meringues from the freezer and leave in refrigerator or a cool room, to allow the cream filling to thaw. Toast the flaked almonds; chop parsley. Put the chicken to cook. Peel and slice the tomatoes for salad, arrange on a serving platter and scatter with finely chopped onion; pour French dressing (see page 83) over ½ hour before serving.

Before the guests arrive   Put the casserole to cook. Cook and drain the rice and rinse in colander with hot water. Leave in the colander over gently steaming water, forking it through occasionally to keep the grains separate. Peel a cucumber, slice thinly and place it in a bowl with a 5-fl oz carton natural yoghurt; chill well before serving with the chicken. Garnish the chicken with the toasted almonds and parsley before serving.

Chicken with curry and lemon

3½–4-lb oven-ready chicken cut into 8 joints
1½ oz seasoned flour
2 level tsps curry powder
2 oz butter
1 tbsp cooking oil
3 small onions
¾ pt chicken stock
1 lemon, cut into thin slices
2 bay leaves

To complete from freezer
1–2 oz flaked toasted almonds
parsley

To make: Wipe the chicken joints. Place the seasoned flour and curry powder in a large polythene bag. Toss the chicken joints in the bag until they are well coated. Melt the butter in a large frying pan, together with the oil. Fry the chicken joints for 10–15 minutes, until really golden brown on all sides; set to one side. Meanwhile, skin and chop the onions. Add to the pan and cook for 5 minutes. Sprinkle any remaining flour into the pan and cook for a further 1 minute, then stir in the stock, lemon slices and bay leaves. Bring to the boil, stirring. Cool quickly.

To pack and freeze: Transfer the chicken to 2 foil freezer trays, or foil-lined casseroles. Pour the sauce over and freeze. Cover with lids or foil. Put in a freezer bag, seal and label.

To thaw: Unwrap and turn into a large casserole or leave in the foil tray. Thaw overnight in the refrigerator and then cook covered at 350°F (mark 4) for 1 hour. Alternatively, cook straight from frozen at 400°F (mark 6) for 1½–2 hours, until the chicken is hot and bubbling.

To serve: Leave in the casserole, or turn it into a hot serving dish; scatter with toasted almonds and freshly chopped parsley.

Meringues Chantilly

3 egg whites
6 oz caster sugar

Filling
⅓ pt double cream, whipped
1 small egg white
½ oz icing sugar
a few drops of vanilla essence

Makes 8

To make: Line 2 baking sheets with non-stick paper. Whisk the egg whites in a deep basin until stiff and dry. Add half the sugar and whisk again until the mixture regains its former stiffness but is no longer opaque. Lastly, fold in the remainder of the caster sugar, using a metal spoon. Using 2 spoons or a large star vegetable nozzle and a forcing bag, shape meringue shells of even size on the baking sheets. Dry off in a very cool oven at 250°F (mark ¼) for several hours, until dry, crisp and very slightly off-white. Cool on a wire rack. Whisk the double cream until it just holds its shape. Whisk the egg white until stiff, and fold into the cream with the icing sugar; stir in the essence lightly. When evenly blended, use to pair the meringue shells together.

To freeze: Place on a flat tray and freeze until the cream is firm. Place in a rigid container such as a cardboard box, separating each with crumpled paper. Overwrap with foil or freezer wrap. Remember that even when frozen, meringues are fragile.

To thaw: Allow to thaw in the refrigerator for about 5 hours, or in a cool room for about 3½ hours.

Meringues Chantilly

13

Individual ratatouilles   *Grissini
Beef goulash with soured cream
*Buttered rice   *Baby Brussels sprouts
Chilled poached pears with Melba sauce
Crème Chantilly

*Starred dishes are recipe suggestions or accompaniments
not suitable for freezing*

Countdown

The day before   Take the goulash from the freezer and loosen the lid. Put in refrigerator.

During the afternoon   Take the ratatouilles from the freezer, loosen wrappings and leave to thaw at room temperature for 4 hours. Prepare the vegetables for the main course. Take Melba sauce and crème Chantilly from freezer, uncover both and leave the Melba sauce to thaw at room temperature, but put the crème Chantilly in the fridge. Remove the pears from the freezer, and leave covered, to thaw at room temperature.

Before guests arrive   Set the oven to 375°F (mark 5) 1½ hours before the guests arrive. Put in the ratatouilles, then the goulash ¼ hour later. Blend the cornflour for thickening the goulash. Cook the rice. Cook the sprouts for the main course and thicken the goulash just before serving.

Individual ratatouilles

¾ **lb courgettes**
1 **lb aubergines**
salt
1 **large red pepper**
1 **small green pepper**
3 **small onions**
½ **lb tomatoes**
4 **tbsps dry white wine**
12 **tbsps olive or cooking oil**
2 **bay leaves**

To complete from freezer
chopped parsley

To make : Follow the same method as that given on page 24 for ratatouille with garlic, but omitting the garlic.

To pack and freeze : Turn the mixture into 4 foil lined dishes (½-pt capacity). Freeze until firm, then remove from the dishes, wrap in foil, overwrap in a freezer bag, seal and label. Return to the freezer.
To thaw : Unwrap and thaw at room temperature for 4 hours.
To use : Cover the dishes and put in the oven at 375°F (mark 4) for 1 hour, forking the mixture through occasionally. If cooking from frozen, put in the oven at the same temperature, but allow 1½ hours. Serve hot or cold; top with some chopped parsley, and serve with grissini (bread sticks).

Beef goulash

1½ **lb best stewing steak**
2 **oz lard**
1 **medium-size onion, skinned and chopped**
1–2 **level tbsps paprika pepper**
1 **level tbsp tomato paste**
just under ½ pt beef stock

To complete from freezer
cornflour
5-fl oz carton soured cream

To make : Wipe, trim and cube beef. Fry it in hot lard until brown on all sides. Add the onion, and cook for 2–3 minutes. Stir in the paprika pepper, tomato paste and beef stock, bring to the boil, then transfer to casserole. Cover and cook towards the centre of the oven at 325°F (mark 3) for 1½ hours. Cool quickly.
To pack and freeze : Transfer to a 4-portion size rigid foil freezer tray, cover, seal, label and freeze.

Or use a foil-lined casserole and remove the foil package when frozen; overwrap with freezer wrap, seal, label and return to the freezer.

To use: Leave goulash in the foil freezer tray or unwrap and turn into a casserole. Thaw overnight in the refrigerator and heat through for about $\frac{3}{4}$ hour in the oven at 375°F (mark 5). Or cook from frozen for $1\frac{1}{4}$ hours at 375°F (mark 5). Blend $\frac{1}{2}$ oz cornflour with 2 tbsps water and stir into the hot goulash to thicken. Serve in the casserole or turn into a hot serving dish. Serve the soured cream separately and spoon on to each helping.

Buttered rice

This recipe is a menu suggestion not suitable for freezing

Allow 2 oz long-grain rice per person. Cook in boiling, salted water until soft – about 15 minutes. Drain and rinse with boiling water.
When dry and fluffy, add a large knob of butter and fork through before turning into a warmed dish.

Poached pears

$\frac{1}{2}$ lb sugar
1 pt water
2 lb pears
lemon juice

To make: Make the syrup by dissolving the sugar in the water, bring to the boil, remove from the heat. Peel, quarter and remove the core from the pears. Poach the pears in the syrup for $1\frac{1}{2}$ minutes.
To pack and freeze: Drain, cool and pack in a rigid polythene container. When cold cover with the syrup, seal, label and freeze.
To use: Allow to thaw at room temperature for 3–4 hours. Spoon into glass dishes, keeping the pears covered in syrup as long as possible to stop discoloration.

Melba sauce

1 lb fresh or frozen raspberries
1 level tbsp cornflour
1 tbsp water
2 oz sugar
finely grated rind and juice of 1 lemon

Makes $\frac{1}{2}$ pt

To make: Hull the raspberries, if necessary. Cook in a covered pan (preferably easy-clean) over gentle heat for about 10–15 minutes until reduced to a pulp. Pass through a nylon sieve into the rinsed pan. Blend the cornflour and water to a cream and stir into the fruit purée; cook over a gentle heat, stirring until the sauce thickens – 1–2 minutes. Add sugar, lemon juice and rind. Allow to become cold.
To pack and freeze: Pour into a $\frac{3}{4}$-pt foil dish or other rigid container. Cover with lid, seal, label and freeze till solid.
To use: Allow 7 hours in refrigerator or $3\frac{1}{2}$ hours at room temperature. Pour the sauce over the chilled poached pears.

Crème Chantilly

$\frac{1}{2}$ pt fresh double cream
1 large egg white (optional)
1 oz icing sugar
few drops of vanilla essence

To make: Whip the cream until thick but not stiff. Fold in the stiffly whisked egg white and icing sugar with a metal spoon. Stir in the vanilla essence lightly.
To pack and freeze: Spoon into a 1-pt foil dish or plastic container. Cover with a lid. Label, seal and freeze until solid.

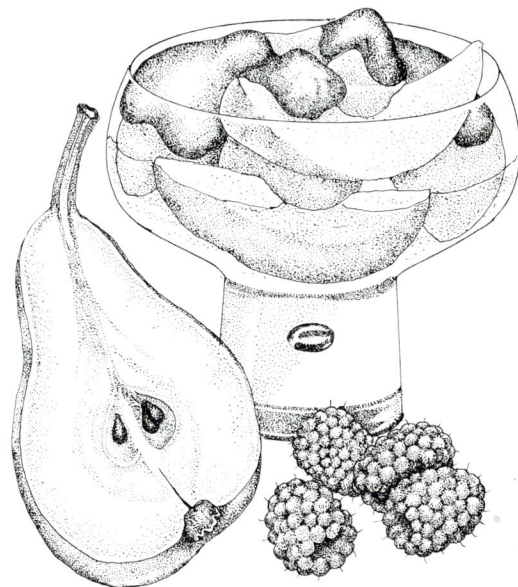

Poached pears with Melba sauce

To thaw: Allow 3 hours in the refrigerator. *To serve:* Place in a glass dish to serve as an accompaniment to the pears. Crème Chantilly can also be used as a filling for a Pavlova or meringue case, and as a topping for trifles (freeze it in whirls – see page 82 – if you like a piped decoration).

FOR FOUR

*Jellied consommé
*Chopped hard-boiled egg and onion   *French bread*

Steak au poivre
*New potatoes   **Green salad*

Cheeseboard

Starred dishes are recipe suggestions or accompaniments not suitable for freezing

Countdown

Early evening   Prepare the potatoes. Make up the green salad. Spoon the jellied consommé into bowls and leave in the fridge until required. Hard-boil and chop 2 eggs and chop 1 onion finely. Place in separate bowls, as accompaniments to the consommé. Arrange the cheeseboard. Cook the potatoes and dress salad (see page 83) just before taking it to table. Start cooking steaks, if fully frozen, before serving the consommé.

Jellied consommé

This recipe is a menu suggestion not suitable for freezing

Empty 2 × 10½-oz cans of consommé into a bowl. Stir in 3 tbsps dry sherry and a squeeze of lemon. Place in refrigerator to chill till jellied – about 2 hours.

Steak au poivre

4 fresh rump steaks (8–10 oz each)
½ oz whole black peppercorns

To complete from freezer
3 oz butter
1 tbsp oil
2 tbsps brandy and ¼ pt double cream (optional)
watercress to garnish

To make: Wipe the steaks. Crush the peppercorns with a rolling pin between sheets of greaseproof paper, or use a pestle and mortar. Coat each side of the steaks lightly with crushed peppercorns, pressing them on lightly with the palm of the hand. *To pack and freeze:* Place on a large sheet of foil with freezer paper or waxed paper between them so that the steaks can be easily separated. Make into a neat parcel. Overwrap, using freezer wrap or foil. Seal and label. Freeze rapidly. *To thaw:* Allow 8–9 hours in the refrigerator or 4 hours at room temperature or cook straight from frozen. *To cook:* Heat 3 oz butter and 1 tbsp oil in a large frying pan. Add the steaks, turn once and cook 5–8 minutes in all if thawed, or 12–15 minutes if frozen. Keep warm on a serving plate. At this stage, if you wish, add 2 tbsps brandy to the juices in the pan, flambé, remove from the heat and stir in ¼ pt double cream. Heat gently, adjust seasoning, pour over the steaks. Garnish with watercress.

Cheeseboard from the freezer

Make a selection from ripe Camembert, Boursin with garlic, or a cream cheese with herbs. Valmeuse or Brie, when ripe, can also be frozen successfully. It is not advisable to freeze hard or blue cheeses, since they develop a crumbly texture which is not suitable for a cheeseboard, though as there is no flavour change they can be used for cooking or for using in sandwiches and salads.

Cheeseboard

To pack and freeze: Wrap closely in a double layer of foil. Overwrap, using a freezer bag. Seal and label the packs. Freeze rapidly.
To thaw: Allow 1–2 days in the refrigerator, then allow a few hours to 'come to' at room temperature before serving.
To serve: Arrange attractively on a board (or platter) and add a touch of garnish-cum-accompaniment, such as small radishes, celery or watercress.

FOR FOUR

Gazpacho

Trout meunière

*Tiny new potatoes   *Courgettes with parsley

Profiteroles   Chocolate and rum sauce

Starred dishes are recipe suggestions or accompaniments
not suitable for freezing

Countdown

The night before   Take the trout from the freezer. Leave wrapped and set on a large plate at room temperature to thaw.
In the morning   Take gazpacho from the freezer, unwrap and place in a bowl. Thaw at room temperature for about 6 hours.
Late afternoon   Prepare selected vegetables. Prepare accompaniments for gazpacho. Put the profiteroles on a baking sheet and refresh in the oven at 350°F (mark 4) for about 10 minutes to crisp up; cool on a wire rack. Fill with whipped cream just before the guests arrive. Chocolate and rum sauce can be unwrapped and heated through straight from the freezer in a double boiler.
When guests arrive   Set vegetables to cook. Cook fish on one side, turn, then serve first course while fish is cooking gently on the other side.

Gazpacho

1 lb tomatoes, skinned, quartered and
    seeded
1 small onion, skinned and quartered
½ cucumber, cut into chunks
4 tbsps wine vinegar
2 tbsps olive oil
2 tbsps red wine

To complete from freezer
1 clove garlic
15-oz can tomato juice
chopped parsley
croûtons
onion
red and green peppers
1 hard-boiled egg

To make : Purée the tomatoes, onion and cucumber in an electric blender. Pour the mixture into a bowl and stir in the vinegar, oil and wine.
To pack and freeze : Line a 2-pt container with a freezer bag and pour in the gazpacho. When it is frozen, remove the bag from the container place inside another freezer bag, seal and label.
To thaw : Leave at room temperature for about 6 hours.
To serve : Add 1 clove of garlic, skinned and crushed, stir in a can of tomato juice, adjust the seasoning and add chopped parsley. Serve with fried bread croûtons, finely chopped onion, chopped red and green peppers and chopped hard-boiled egg.

Trout meunière

4 rainbow trout, about 8 oz each *(see below)*

To complete for serving
2 oz seasoned flour
2–3 oz butter
2 tbsps oil
2 lemons
watercress

To prepare and pack : Choose only very fresh fish, not previously frozen; ask the fishmonger to clean them for you. Wipe each fish with a damp cloth. Either wrap in freezer film, then in foil *or* treat as follows : place the fish in a shallow bowl of water with plenty of ice cubes. Turn each fish over and

over in the water, then place on a wire tray in the freezer till frozen. Repeat, immersing the fish in more iced water, and then returning it to the freezer; do this 5 times altogether. Gradually a coat of ice builds up which prevents evaporation. Overwrap in double foil and then place in a freezer bag. Seal and label.
To thaw : Set the fish on a large plate. Leave overnight in the refrigerator, still in the wrappings. Only if time is very short should you thaw the fish at room temperature, in a bowl of cold water, or cook them from frozen; in this case, allow ½ hour cooking time, and use a large pan, as the fish, being rigid, will not 'give' to the shape of a smaller pan.
To cook : Place the fish on absorbent kitchen paper and toss in 2 oz seasoned flour. Fry in 2–3 oz hot butter and 2 tbsps oil, allowing 5 minutes on each side. Turn the fish carefully. Lift them out on to a serving plate. Add the juice of 1 lemon to the pan juices, allow to boil, then pour over the fish immediately.
To serve : Garnish with lemon wedges and sprigs of watercress.

Profiteroles

2 oz butter
¼ pt water
2½ oz plain flour
2 eggs

To complete from freezer
½ pt double cream, whipped
Chocolate and rum sauce *(see below)*

Makes 20

To make : Melt the butter in the water and bring to the boil. Remove from the heat and add the flour at once. Beat until the paste is smooth and forms a ball in the centre of the pan. (Take care not to overbeat or the mixture will become fatty.) Allow to cool slightly, then beat in the eggs gradually, adding just enough to give a smooth mixture of piping consistency. Using a large plain nozzle, pipe the paste on to greased baking sheets in small dots. Bake in the oven at 425°F (mark 7) for about 20–25 minutes. Allow to cool and make a small hole for steam to escape.
To pack and freeze : Place the buns on foil-lined

baking sheets and open freeze until solid. Remove carefully from the sheets and pack in freezer bags or heavy duty foil. Overwrap, seal, label and return to the freezer.

To use: Thaw wrapped at room temperature for about 1 hour then remove the wrappings, place the buns on baking sheets and refresh in the oven at 350°F (mark 4) for about 5 minutes. Alternatively, unwrap and refresh from frozen at 350°F (mark 4) for about 10 minutes. Cool on a wire rack. Make a small hole in each and fill with whipped cream. Pile into a pyramid on a serving plate and spoon a little Chocolate and rum sauce over. Serve any remaining sauce separately.

Chocolate and rum sauce

8 oz plain chocolate
2 oz unsalted butter
4 tbsps milk
2 tbsps rum

Makes ½ pt

To make: Break the chocolate into small pieces and place in a bowl over a pan of hot water. Add the butter and milk. Stir occasionally until glossy and smooth. Stir in the rum.

To pack and freeze: Pour into a foil container or plastic tub. Overwrap, label and seal. Freeze until solid.

To use: Unwrap then place the sauce in a double boiler and cook gently, stirring, till the sauce is smooth. Pour over the profiteroles, or serve separately.

Notes The sauce separates out during freezing, but returns to a good consistency on being reheated.

If you should have a quantity of sauce left over after the meal, cool it, then pack, seal and label and return it to the freezer.

Profiteroles

FOR SIX

Herb pâté   Fresh toast

Sherry chicken
*Broccoli with almonds   Duchesse potatoes

Gooseberry sorbet

*Starred dishes are recipe suggestions or accompaniments
not suitable for freezing*

Countdown

The night before   Remove the pâté from the freezer and allow to thaw at room temperature.
Early evening   Remove the chicken from the freezer and begin to cook according to instructions. Prepare potatoes and ingredients for broccoli with almonds. Place the sorbet in the refrigerator.
Before the guests arrive   Complete the chicken dish. Cook potatoes and broccoli.

Make toast just before serving the pâté. Spoon the sorbet into glasses just before serving.

Herb pâté

1 lb pig's liver
¾ lb streaky bacon rashers or pieces
1 small onion, skinned and finely chopped
pinch of ground black pepper
¼ level tsp ground mace
pinch of grated nutmeg
¼ level tsp mixed dried herbs
1 small egg, beaten

To make: Wash the liver well; cut the rind from the bacon and remove any excess fat. Fry the bacon rind and fat trimmings gently until the fat runs. Discard the rinds and pieces of fat and add the onion to the pan. Fry the onion until soft and transparent, for about 4 minutes. Mince the liver and bacon three times to give a fine smooth texture. Put in a large bowl with the onion, pepper, mace, nutmeg and mixed herbs. Add the egg and mix the ingredients together until well blended. Spoon the mixture into an 8½-in by 4½-in loaf tin (top measurement) lined with greaseproof paper. Cover the top with aluminium foil and place in a roasting tin containing 2 in water. Cook in the oven at 350°F (mark 4) for 1¾ hours or until firm. Leave in the tin with a weight on top to become cold.

To pack and freeze: Turn out, remove the paper, wrap in foil and overwrap with freezer wrap. Seal, label and freeze.

To use: Remove freezer wrap, loosen foil wrapping and thaw for up to 24 hours at room temperature or 48 hours in the refrigerator. Serve with hot toast.

Sherry chicken

6 chicken portions
¼ pt dry sherry
2 level tsps caster sugar
3 tbsps cooking oil
3 oz mushrooms, sliced
1 level tsp dried oregano
1 pt chicken stock
salt and pepper

To complete from freezer
2 level tbsps cornflour

To make: Marinate the chicken pieces in the sherry and sugar for 30 minutes. Heat the oil in a frying pan and fry the chicken until brown on all sides. Add the mushrooms, oregano, marinade and stock. Bring to the boil, season with salt and pepper, cover and simmer for 1 hour. Cool rapidly and skim off any excess fat.

To pack and freeze: Spoon into a foil dish, cover lightly with the lid and freeze. Then seal the lid tightly, label and return to the freezer. Or line a casserole with foil, spoon in the chicken and freeze. When solid, remove the foil covered block, wrap the foil round, overwrap with freezer wrap, seal and label. Return to the freezer.

To use: Unwrap and cook in the foil dish, covered with foil, or in a covered casserole. Cook from frozen in the oven at 350°F (mark 4) for about 2 hours. Place the chicken pieces on a warm serving dish and keep hot. Blend the cornflour with a little of the cooking liquid, add to the remaining liquid. Bring to the boil and cook until thickened. Pour over the chicken.

Duchesse potatoes

3 lb old potatoes, peeled and boiled
1 oz butter
½ egg
½ level tsp salt
freshly ground black pepper
¼ level tsp grated nutmeg

To complete from freezer
1 egg, beaten

To make: Mash the potatoes, add the remaining ingredients (no milk) and beat well. Line a baking sheet with non-stick paper. Using a forcing bag fitted with a large star nozzle, pipe on to the sheet about 12 raised pyramids of potato, with a base of about 2 in.

To pack and freeze: Freeze uncovered·until firm. Remove from the freezer. Slide the potato pyramids off the tray on to a foil plate or container. Cover with freezer film, seal and label.

To use: Grease some baking sheets or line them with kitchen foil, transfer the frozen duchesse potato portions to the baking sheets, brush lightly with egg glaze and put into a cold oven. Cook at 350°F (mark 4) for 20–30 minutes, or until heated and lightly browned.

Note Milk has been omitted from the recipe because it is found that when defrosted, the potato mixture 'weeps' slightly, owing to the large amount of water present.

Broccoli with almonds

This recipe is a menu suggestion not suitable for freezing

1½ lb frozen broccoli
2 oz butter
1 onion, skinned and finely chopped
1 clove garlic, skinned and crushed with a little salt
salt and freshly ground black pepper
½ tsp lemon juice
1½ oz almonds, blanched, shredded and toasted

Cook the broccoli in boiling salted water until just tender, drain it well and arrange in a warm serving dish. Keep it warm. Melt the butter and gently sauté the onion and garlic until transparent and add salt, pepper and lemon juice. Spoon the onion and butter over the broccoli and scatter with the almonds.

Gooseberry sorbet

2 lb gooseberries, topped and tailed
¼ pt water
8 oz caster sugar
2 tbsps lemon juice
edible green colouring (optional)
2 egg whites

Gooseberry sorbet

To make: Place the gooseberries in a saucepan with the water, bring to the boil, cover and cook until soft. Stir in the sugar and cool. Put through a sieve into a freezer container, stir in the lemon juice and colouring if desired. Freeze until mushy. Whisk the egg whites until stiff and fold into the half-frozen mixture. Turn into a container and freeze.
Note When fresh fruit is not available use a can but omit the water and sugar, or use frozen fruit.
To pack: Seal and label.
To use: Allow to 'come to' in the refrigerator for about 2 hours before serving, spooned into glass dishes.

FOR SIX

Coquilles St Jacques

Noisettes of lamb
*Buttered new potatoes   *Tossed green salad

Pavlova with peaches

Starred dishes are recipe suggestions or accompaniments not suitable for freezing

Countdown

In the morning   Take the noisettes and the coquilles from the freezer and allow to thaw as directed in the recipes.

Coquilles St Jacques

Late afternoon Prepare the vegetables. Rinse the salad ingredients, drain thoroughly, then put them in to a large polythene bag in the refrigerator to crisp. Fry the croûtes. Whip the cream for the Pavlova and drain the sliced peaches. Fill the Pavlova case with cream and peaches and decorate with melted chocolate.

Before guests arrive Arrange the salad and make up the dressing (see page 83). Put the coquilles to cook. Cook potatoes and keep hot. Set the noisettes to cook in the grill pan.

Coquilles St Jacques

½ lb shelled fresh scallops
¼ pt dry white wine
¼ of a small onion, skinned
sprig of parsley
1 bay leaf
1 oz butter
2 oz button mushrooms, sliced
6 small natural scallop shells, washed
duchesse potato mix *(see page 20)*

For the sauce
2 oz butter
2 oz plain flour
¾ pt milk
2 oz grated cheese
seasoning

To complete from freezer
wedges of lemon, *or*
parsley sprigs

To make : Rinse, and slice the white parts of each scallop into 4; leave the coral whole. Place in a pan with the wine, onion, parsley and bay leaf. Bring to the boil and simmer for 5 minutes. Drain, keeping the strained liquor to one side. Melt the butter and sauté the mushrooms for 5 minutes. Prepare the sauce: melt the butter in a pan, add the flour and cook over a gentle heat without browning for 1 minute; stir in the strained liquor and the milk. Return the pan to the heat, bring to the boil, stirring till the sauce is smooth and thickens. Cool slightly, then stir in the sautéd mushrooms, cheese and scallops, with seasoning to taste.

To pack and freeze : Divide the mixture between the scallop shells (or 6 individual ovenproof dishes). Pipe duchesse potato mixture round the shells and brush with beaten egg (though this can be done when the scallops are taken from the freezer before cooking). Freeze uncovered, so as not to damage the piped potato. Place inside 2 freezer bags. Seal and label.

To thaw : Remove the seal and packaging, and if a potato border was not added before freezing, pipe it on now. Set the coquilles on a baking sheet, loosely covered so as not to damage the potato; leave in the refrigerator for about 8 hours.

To serve : If taken straight from freezer, remove the outer wrapping; place the coquilles on a baking sheet. Cover loosely with foil and place in the oven at 425°F (mark 7). If thawed, reheat for 15 minutes; otherwise, allow 50–60 minutes. Uncover, brush with beaten egg and cook for a further 10–15 minutes, until golden. Garnish with wedges of lemon or parsley sprigs.

Noisettes of lamb

**2 best ends of neck of lamb (6 noisettes
  each)**

To complete from freezer
maître d'hôtel butter
croûtes
parsley

To make : Ask your butcher to bone and roll the best ends. Wipe the meat with a clean, damp cloth, then cut each best end into 6 noisettes, cutting between the strings.

To pack and freeze : Place pieces of freezer paper or waxed paper between the noisettes so that they will separate easily before cooking. Cover in a double layer of foil. Overwrap, using freezer bag. Seal and label.

To thaw : Place on a large plate and allow 12 hours in the refrigerator, or 5 hours at room temperature (or cook from frozen).

To cook : Unwrap, then season to taste and dot with butter. Whether frozen or already thawed, brown them quickly on both sides under a hot grill, then reduce the heat and cook for a further 10–15 minutes if thawed, or 30 minutes if frozen; baste frequently. They should still be quite pink in the centre.

To serve : Place some maître d'hôtel butter on each noisette (see page 81) and set on a croûte of fried bread. Garnish with sprigs of parsley.

Pavlova with peaches

3 egg whites
6 oz caster sugar
$\frac{1}{2}$ tsp vanilla essence
$\frac{1}{2}$ tsp vinegar
2 level tsps cornflour

To complete from freezer
$\frac{1}{2}$ pt double cream, whipped
16-oz can sliced peaches
2 oz plain chocolate

To make : Draw an 8-in circle on a sheet of non-stick paper and place the paper on a baking sheet. Beat the egg whites till very stiff, then beat in the sugar half at a time. Beat in the vanilla essence, vinegar and cornflour. Spread the meringue mixture on the paper over the circle, piling it up round the edges to form a case. Bake in the centre of the oven at 300°F (mark 1–2) for about 1 hour, till firm. Leave to cool, then carefully remove paper.

To pack and freeze : Use a cardboard box or plastic box, or place on a cardboard base or baking sheet and overwrap with freezer film. Seal and label. Place in the freezer on top of other items, as a Pavlova is still quite fragile, even frozen. Freeze until solid.

To serve : The cake is ready to serve straight from the freezer. Unwrap and fill with the whipped cream and drained peaches. Melt the chocolate, spoon it into a forcing bag fitted with a No. 2 trellis pipe and run it over the peaches in petal shapes. Place the Pavlova in the refrigerator until required.

FOR SIX

Ratatouille with garlic

Steak, kidney and oyster pie
*Buttered spring greens   *Creamy buttered potatoes

Caramel custard
Sponge fingers

*Starred dishes are recipe suggestions or accompaniments
not suitable for freezing*

Countdown

The night before   Take the ratatouille and the steak, kidney and oyster pie from the freezer. Remove seals and covers and set in the refrigerator to thaw.

Early afternoon   Cook caramel custard; leave to cool, then set in refrigerator to chill.

Late afternoon   Remove sponge fingers from the freezer and leave to thaw at room temperature. Prepare the vegetables. Set the ratatouille to cook. Peel and cut the tomatoes into small pieces. Chop the parsley. Add tomatoes to the ratatouille.

Before guests arrive   Set the oven to 450°F (mark 8) and put the pie in when the oven is hot.

When the guests arrive, cook the potatoes then drain and mash. Turn down the oven to 325°F (mark 3). Put greens to cook while eating ratatouille. Drain thoroughly before serving and pour a little melted butter over them before taking them to the table.

Ratatouille with garlic

1¼ lb aubergines
1 lb courgettes
2 level tbsps salt
1 red pepper
1 green pepper
1 large onion
¼ pt plus 2 tbsps oil
4 tbsps dry white wine

To complete from freezer
1 lb tomatoes
1 clove garlic
parsley
grated Parmesan cheese

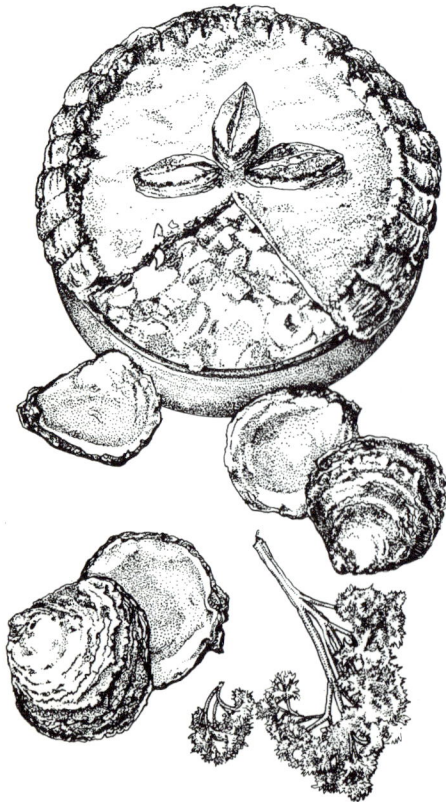

Steak, kidney and oyster pie

To make: Wash, dry and thinly slice the aubergines and courgettes, using a stainless steel knife. Spread out on a large tray or plate, sprinkle with salt and leave for 1 hour. Drain away any excess liquid, then dry the aubergines and courgettes with absorbent paper. Meanwhile cut a slice from the stalk end of the peppers; scoop out and discard the seeds and pith. Cut the flesh into rings. Skin and slice the onion thinly. Heat half the oil in a pan. Add the peppers and onion, then cook for 2–3 minutes. Lift out with a draining spoon into a rigid foil container. Add the remaining oil to the pan and cook aubergines and courgettes for 5–8 minutes, stirring occasionally. Lift from the pan as for peppers, add to the container and spoon in the wine, allow to cool.

To pack and freeze: Cover with the lid or a double layer of foil and freeze until firm; overwrap with freezer wrap, seal and label.

To use: Thaw in the refrigerator overnight. Cover and put in the oven at 400°F (mark 6) for ½ hour. Remove the cover. Add 1 lb fresh firm tomatoes, skinned and cut into 6 or 8 pieces, and 1 clove of garlic, skinned and crushed. Season to taste and stir lightly. Cook uncovered for a further 30–40 minutes. Sprinkle with chopped parsley and serve hot or cold. Serve grated Parmesan cheese separately.

Steak, kidney and oyster pie

2 lb best stewing steak
8 oz ox kidney
1 oz seasoned flour
2 oz lard
2 medium-size onions, skinned and chopped
2 level tbsps tomato paste
½ pt beef stock
1 bay leaf
seasoning
6 fresh oysters
¾ lb ready-made puff pastry

To complete from freezer
sprig of parsley

To make: Wipe, trim and cube the beef. Rinse and trim the kidney and cut into small pieces. Toss meat and kidney in seasoned flour in a large polythene bag until well coated. Fry quickly in melted lard until brown on all sides. Add onion and cook for 3 minutes. Stir in tomato paste blended with stock; add the bay leaf and a little seasoning. Pour into an ovenproof casserole, cover and cook in oven at 325°F (mark 3) for 1¾ hours, until tender. Cool quickly. Remove the bay leaf, then place pie filling in a 2½-pt pie dish; drop in the fresh oysters. Roll out pastry and use to cover the pie dish. Decorate with pastry leaves. Do not make an air vent in the pastry.

To pack and freeze: Put in a freezer bag and freeze, taking care not to damage pastry. Remove from freezer and overwrap with foil, or place inside another freezer bag. Seal well and label.

To use: Remove overwrapping and loosen seal, then set pie in refrigerator on a baking sheet for 12–16 hours. Remove freezer bag and bake pie in the oven at 450°F (mark 8) for 20 minutes, then make an air vent in the centre of the top, and return the pie to the oven. Reduce the temperature to 325°F (mark 3) and bake for a further 20 minutes. Cover the top of the pie with foil if there is a danger of over-browning. Garnish with a sprig of parsley.

Caramel custard

4½ oz sugar
¼ pt water
1 pt milk
4 large eggs

To complete from freezer
½ pt single cream

To make: Put 4 oz of the sugar and the water in a small pan and dissolve the sugar slowly; bring to the boil and boil without stirring until it caramelises, ie becomes a rich golden brown colour. Pour the caramel into a 6-in (2-pt) soufflé dish, turning the dish until the bottom is completely covered. Warm the milk, pour on to the lightly beaten eggs and remaining sugar and strain over the cooked caramel.

To pack and freeze: Allow to cool completely, cover with a foil cap and freeze. When quite frozen, overwrap in a freezer bag. Seal and label.

To use: Remove coverings. Set frozen custard in a roasting tin of warm water, which should come half-way up the sides of the dish. Put in the oven at 375°F (mark 5) and cook for about 1¼ hours until set. Serve cold – leave in dish in the refrigerator until cold before unmoulding. Serve with pouring cream.

Sponge drops and fingers

3 oz caster sugar
3 large eggs
vanilla essence
3 oz plain flour

To make: Grease a baking sheet, or line it with non-stick paper. Whisk the sugar and eggs together until really thick and creamy; add a few drops of vanilla essence. Sift in the flour, folding in a little at a time. Spoon the mixture into a forcing bag fitted with ½-in plain nozzle. Pipe small drops or fingers well apart on the baking sheet and bake in the oven at 425°F (mark 7) for 7–10 minutes, until evenly but lightly browned. Use a palette knife to lift on to a wire rack to cool.

To pack and freeze: Open freeze, spread on baking sheets. Place carefully in a foil container, overwrap in freezer wrap. Seal, label and return to the freezer.

To use: Remove from the freezer and allow to thaw at room temperature for 2 hours.

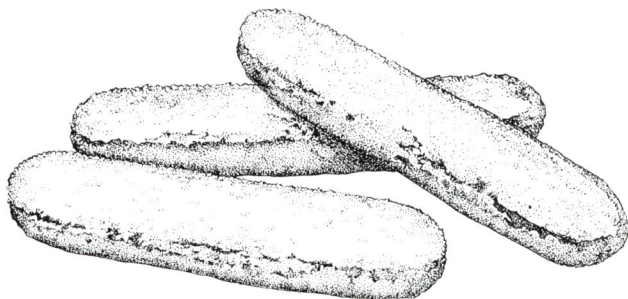

Sponge fingers

Smoked salmon mousse

Spicy pepper casserole
*Buttered new potatoes   *Courgettes

Rum and raisin ice cream

*Starred dishes are recipe suggestions or accompaniments
not suitable for freezing*

Countdown

The night before   Remove the mousse from the freezer and allow to thaw overnight in the refrigerator.

Early evening   Prepare the potatoes and courgettes for cooking. Remove casserole from freezer, place in a casserole dish and begin to cook according to instructions.

Before the guests arrive   Place the mousse on a serving plate and decorate. Cook the vegetables. Stir the yoghurt into the casserole. Place the ice cream in the refrigerator to 'come to'. Spoon into glasses just before serving.

Smoked salmon mousse

½ pt milk
1 small carrot, skinned and chopped
1 small onion, skinned and halved
3 parsley stalks
6 peppercorns
1½ oz butter
1 oz flour
4 oz smoked salmon trimmings
½ oz gelatine
¾ pt chicken stock
juice of ½ lemon
2 level tbsps mayonnaise
¼ pt double cream
salt and freshly ground black pepper

To make: Pour the milk into a saucepan, add the carrot, onion, parsley stalks and peppercorns. Bring to the boil, remove from the heat and allow to infuse for 15 minutes. Melt the butter in another saucepan, stir in the flour and cook for 1–2 minutes without browning. Off the heat, add the strained milk gradually, then bring to the boil and cook,

stirring, for ½ minute. Cover the sauce with greaseproof paper and let it cool.

Chop the smoked salmon coarsely. Dissolve the gelatine in some of the stock, add the rest of the stock and allow to cool. Fold this into the white sauce. Add the smoked salmon, lemon juice and mayonnaise. Beat the cream lightly and fold it into the sauce mixture. Pour the mixture into a 6-in soufflé dish or 6 individual cocotte dishes. Leave to set.

To pack and freeze: Cover with a foil cap and overwrap with freezer wrap, label and freeze.

To use: Allow to thaw overnight in the refrigerator. Dip the container in hot water and invert on to a serving plate. Garnish with lemon. Serve with Melba toast.

Spicy pepper casserole

2 tbsps oil
2 onions, skinned and sliced
1 red pepper, seeded and sliced
1 green pepper, seeded and sliced
2 lb shoulder pork, trimmed
2 level tbsps flour
salt and pepper
1 level tsp paprika
1 pt stock
1 level tbsp tomato paste

To complete from freezer
5-fl oz carton natural yoghurt

To make: Heat the oil in a large saucepan and gently fry the onion until soft. Transfer to a casserole, then fry the peppers and add to the dish. Cut up the meat into 1-in cubes and toss in flour seasoned with salt, pepper and paprika. Brown in the reheated oil. Add the stock and tomato paste;

add the cool mixture, allow to freeze until firm, then remove from dish and overwrap with freezer wrap. Alternatively, a large rigid foil container with lid can be used. Seal well and label.

To use: Remove from freezer and reheat from frozen at 425°F (mark 7) for about 1 hour. If mixture is wrapped in foil, unwrap and return it to original dish before reheating. Just before serving, pour over the yoghurt.

Rum and raisin ice cream

4 oz raisins
4 tbsps rum
4 eggs
4 oz icing sugar
½ pt double cream

To make: Soak the raisins in rum for about 30 minutes. Separate the eggs and whisk the yolks and sugar together until thick and fluffy. Beat the cream until thick and fold into the egg mixture. Add the rum-soaked raisins. Whisk the egg whites until stiff and fold into the mixture. Pour into a polythene container.

To pack and freeze: Freeze for about 1 hour, then stir the mixture to redistribute the raisins. Seal the container, label, and return to the freezer.

To use: Place the container in the refrigerator for half an hour before spooning the ice cream into individual serving dishes.

Spicy pepper casserole

bring carefully to the boil, stirring. Add to the casserole, cover, and cook at 350°F (mark 4) in oven centre for about 1½ hours. Cool quickly.

To pack and freeze: Line a clean casserole with foil,

FOR SIX

Smoked trout or mackerel

Chicken and lemon pie
*Leaf spinach   *Lyonnaise potatoes
*Liqueur cherry brûlée

Starred dishes are recipe suggestions or accompaniments not suitable for freezing

Countdown

First thing in the morning   Take the trout or mackerel and chicken pie from the freezer. Loosen the wrappers from the fish but leave in the freezer bag. Remove the foil wrappings from the pie and set it on a baking sheet in a large

freezer bag. Set both low down in the refrigerator. Leave for about 10–12 hours to thaw.
Noon Prepare the brûlée and put to chill.
Late afternoon Prepare the vegetables. Make sauces for the fish.
Two hours before the meal Finish the cherry brûlée and chill until required. Skin the fish, leaving on the heads and tails if serving whole. Wash the lettuce for garnish, to arrange at the last minute; cut the lemons. Cut the bread, butter it and arrange on a plate. Put the chicken pie in the oven (first removing the polythene bag). Cook the lyonnaise potatoes and spinach. The smoked fish can be taken to the table before the guests are seated.

Smoked trout or mackerel

Allow 1 fish or 1 large fillet per person

To serve from freezer
lettuce
lemon wedges
brown bread

Smoked trout

To pack: Wrap each fish in freezer paper or waxed paper so that they can be easily separated. Then overwrap in foil.
To freeze: Overwrap all fish, using a freezer bag. Seal, label and freeze rapidly.
To use: Allow 10–12 hours in the refrigerator or 8 hours at room temperature. When thawed, lift off the skin from the body of the fish, leaving head and tail intact if serving whole. Serve on lettuce, with lemon wedges and either horseradish cream or mild mustard sauce.

Horseradish cream

Fold 1 level tsp creamed horseradish into $\frac{1}{4}$ pt double cream.

Mustard sauce

Add 1 level tsp made mild mustard and a squeeze of lemon juice to 4 tbsps thick mayonnaise.

Chicken and lemon pie

1 pt milk
juice and thinly pared rind of 1 lemon
2 oz butter
2 oz plain flour
$\frac{1}{2}$ pt chicken stock
$\frac{1}{2}$ level tsp sugar
seasoning
1 lb cooked chicken meat
11-oz can sweet corn kernels
8 oz shortcrust pastry (8 oz flour, etc)

To make: Infuse the lemon rind in the milk in a pan over a gentle heat for 10 minutes. Remove from the heat and cool; discard the lemon rind. Melt the butter in a saucepan, stir in the flour and cook this roux over a low heat for 1 minute. Gradually add the flavoured milk and stock, stirring well. Return to the heat and stir whilst bringing to the boil; cook for a few minutes. Add the sugar, lemon juice and seasoning to taste; cover and cool, stirring occasionally. Cut the chicken meat into small pieces. Add the drained sweet corn and turn the mixture into a $2\frac{1}{2}$-pt pie dish. Spoon the lemon sauce over the chicken and cover with a pastry lid. Decorate with pastry trimmings. (Don't make a hole in the centre of the pastry.)
To freeze and pack: Freeze uncovered. When firm, overwrap in double foil or 2 freezer bags. Seal and label.

To thaw: Remove all wrapping. Place the pie on a baking sheet and cover loosely with a freezer bag. Leave in the refrigerator for 10–12 hours.

To cook: Remove the freezer bag. Brush the pastry with beaten egg and cook in the oven at 400°F (mark 6) for 15 minutes. Make a hole in the centre of the pastry, return the pie to the oven and continue to cook for a further $\frac{1}{2}-\frac{3}{4}$ hour, until the pastry is brown and the filling bubbling. Cover loosely with foil if it shows signs of over-browning.

Lyonnaise potatoes

This recipe is a menu suggestion not suitable for freezing

2 lb potatoes, peeled
salt
6 tbsps oil
1 lb onions, skinned and sliced
chopped parsley

Boil the potatoes in salted water until just cooked, drain well. Cut into $\frac{1}{4}$-in slices. Fry in the fat until crisp and brown on both sides. Drain on absorbent paper. Keep warm.

Fry the onions in the frying pan until golden brown – about 10 minutes. Serve in layers with the potatoes and sprinkle with chopped parsley.

Liqueur cherry brûlée

This recipe is a menu suggestion not suitable for freezing

1 lb frozen unsweetened cherries, thawed and stoned
8 oz caster sugar
2 tbsps Kirsch
$\frac{1}{4}$ pt double cream
$\frac{1}{4}$ pt single cream

Place the cherries in the base of a flameproof serving dish. Sprinkle with 2 oz sugar and the Kirsch. Whip the creams together, spread them on top of the cherries and chill for 2 hours. Sprinkle 6 oz sugar over the cream, place the dish under a preheated grill and allow the sugar to caramelise. Chill the brûlée until required.

FOR SIX

Prawn and cod bisque

Breaded veal escalopes

Sauce aurore

*Fluffy whipped potatoes   *Petits pois

Pineapple creams

Starred dishes are recipe suggestions or accompaniments not suitable for freezing

Countdown

The day before Take the bisque from the freezer and leave wrapped, but place in a bowl in the refrigerator. (In an emergency, heat it on the actual day, straight from the freezer.)

During the morning Take the escalopes and sauce from the freezer and place, still wrapped, in the refrigerator.

Late afternoon Take the pineapple creams from the freezer, unwrap and leave in the refrigerator to thaw. Prepare vegetables. Whisk the sauce aurore, adding extra lemon juice to help whisk it; pour into a serving dish. Decorate the pineapple creams.

When the guests arrive Start cooking the potatoes. Heat the soup, add cream and garnishes.

Cream potatoes, season and turn into warmed serving dish; cover with lid and keep warm. Start cooking the escalopes in melted butter; if necessary, arrange on a serving platter, cover with foil and keep warm whilst cooking the remainder. Set peas to cook while the soup is being eaten. Garnish the escalopes with lemon and take to the table, with the sauce in a separate dish.

Prawn and cod bisque

4 oz cod fillet, skinned
1 small onion, skinned and chopped
1 medium-size carrot, pared and sliced
4 oz button mushrooms, halved
2 oz butter
2 pt fish or chicken stock
juice and $\frac{1}{2}$ level tsp grated rind from a
    small lemon
2 tbsps white wine, optional
seasoning

To complete from freezer
4 oz fresh or frozen peeled prawns
$\frac{1}{2}$ pt single cream
7-oz can sweet corn kernels *or* 2 oz diced
    cucumber
parsley

Makes $2\frac{3}{4}$ pt

To make: Cut the cod into 1-in cubes. Add the onion, carrot and mushrooms to the melted butter in a pan; sauté gently for 5 minutes. Add the cubes of cod and cook for a further 5 minutes. Purée in an electric blender with some of the stock, the lemon juice, lemon rind and wine; season to taste, then cool quickly.
To pack and freeze: Pour into a freezer bag placed in a rigid container. Freeze until firm, remove from the container and place inside another freezer bag. Seal and label.
To thaw: Allow 24 hours in the refrigerator.
To serve: Pour the thawed soup into a saucepan and heat through very gently. Alternatively, unwrap the frozen soup and heat very slowly in a double saucepan. When the soup has just come to the boil, remove from the heat. Cool slightly, then stir in the prawns and cream and adjust the seasoning. Add the drained canned sweet corn kernels or diced cucumber and some chopped parsley. Reheat, but don't boil.

Breaded veal escalopes

6 veal escalopes (4 oz each)
2 oz seasoned flour
2 eggs, beaten
6 oz fresh breadcrumbs

To complete from freezer
butter for frying
lemon wedges

To make: Ask your butcher to beat the escalopes thin. Coat both sides with flour on a sheet of greaseproof paper. Dip each escalope in beaten egg and toss in breadcrumbs. Use the palm of the hand to pat the crumbs well on to the surface of the escalopes.
To pack and freeze: Place on a large sheet of foil with layers of freezer paper or waxed paper between so that they will separate easily. Wrap up securely in a parcel. Overwrap using a freezer bag. Seal, label and freeze rapidly.
To thaw: Allow 8 hours in the refrigerator or cook from frozen.
To cook: Melt 2–3 oz butter in a frying pan. Cook the escalopes 2 at a time, allowing about 5 minutes in all for thawed meat, or about 8–10 minutes if cooking straight from the freezer. Add more butter as required. Garnish with lemon wedges.

Sauce aurore

2 × 5-fl oz cartons soured cream
1–2 tbsps tomato ketchup
grated lemon rind
4 tbsps lemon juice

To complete from freezer
1 tbsp lemon juice
chopped chives, optional

Makes $\frac{1}{2}$ pt

To make: Combine the soured cream with the ketchup. Finely grate a little rind from the lemon and add to the soured cream, with 4 tbsps strained lemon juice.
To pack and freeze: Spoon into a $\frac{3}{4}$-pt foil dish or plastic container. Cover, label and freeze.
To thaw: Allow 8 hours in refrigerator or 3 hours at room temperature.
To serve: Whisk well, adding a further 1 tbsp lemon juice. Sprinkle with chopped chives before serving.

Pineapple creams

15-oz can crushed pineapple
juice of 1 lemon
2 oz caster sugar
½ pt double cream, whipped
4 egg whites, stiffly whisked

To complete from freezer
whipped cream
crystallised violets

To make: Drain off any juice from the crushed pineapple, then purée the fruit in an electric blender to give ½ pt purée. Add the lemon juice and sugar. Fold in the whipped cream, followed by the stiffly whisked egg whites.
To pack and freeze: Turn the mixture into 6 small individual soufflé dishes. Freeze until firm, cover the tops with foil, then overwrap in a freezer bag, seal and label. Return them to the freezer.
To thaw: Uncover, remove the foil and thaw in the refrigerator for 8 hours, or at room temperature for 3 hours.
To serve: Decorate with whipped cream and crystallised violets.

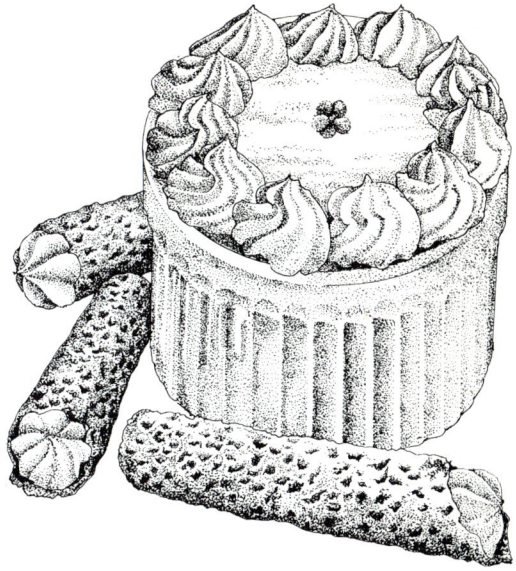

Pineapple creams

FOR SIX

Salmon mousse   Fresh toast

Boeuf en croûte

*Potatoes with chives   *Tomato, chicory and lettuce salad

Blackcurrant and lemon sorbets

*Starred dishes are recipe suggestions or accompaniments
not suitable for freezing*

Countdown

Early the previous day Take the boeuf en croûte out of the freezer; leave wrappings in place but loosen the seal. Set on a plate in the refrigerator for about 36 hours before cooking so that it will be totally thawed.
Late afternoon on the day Take individual salmon mousses from the freezer, unwrap and leave at room temperature to thaw. Prepare potatoes cut into small pieces and parboil, drain.

Prepare ingredients for salad. Pour dressing ingredients into jar (see page 83), but don't shake up. Garnish the salmon mousses. Cut bread ready for toasting. Arrange salad in bowl.
When guests arrive Place potatoes in hot fat and put boeuf en croûte in oven. Drain sauté potatoes, keep hot; sprinkle with chopped chives before serving. Dress salad before taking it to table and take sorbets from freezer just before serving main course.

Salmon mousse

2 × 7½-oz cans red salmon
juice and grated rind of 1 lemon
1 oz gelatine
¼ pt water
½ pt double cream, whipped
salt and freshly milled pepper

To complete from freezer
cucumber
parsley

To make: Remove any dark skin from the salmon
and mash, with the juices, on a large plate (even
better, purée, using an electric blender). Turn it
into a bowl. Add the lemon juice and grated rind.
Dissolve the gelatine in the water in a small bowl
set in a pan of gently bubbling water. When cool
but still liquid, stir into the fish mixture. Fold in
the cream, which should just hold its shape. Adjust
the seasoning.
To pack and freeze: Put into 6 individual ramekin
dishes; cover with foil caps. Place on a baking sheet
and freeze. Overwrap with a double layer of foil,
seal, label and return to freezer.
To thaw: Place in the refrigerator to thaw
overnight or allow 4 hours at room temperature.
To serve: Garnish with cucumber slices or twists
and some chopped parsley.

Boeuf en croûte

2 lb fillet of beef
2 rashers of bacon, rinded
salt and pepper
½ lb button mushrooms, sliced
2 oz butter
¾ lb ready-made puff pastry

To complete from freezer
1 egg, beaten
watercress, to garnish

To make: Place the beef in a roasting tin with the
bacon rashers laid across the surface. Cook in the
oven at 400°F (mark 6) for 15 minutes. Take from
the oven and remove the bacon; leave to cool
rapidly. Sauté the seasoned mushrooms in butter
for 3–4 minutes; drain. Either sieve the mushrooms or chop very finely. Leave to become cold.
Roll out the pastry into an oblong 14 in long, and
the width of the fillet plus 3 in. Cover centre area of
pastry with mushrooms, then place the beef on
top. Damp the edges of the pastry, then bring up
the edges over the ends of the fillet to form a neat
parcel – make sure that the edges are well sealed.
Invert so that the sealed edges are underneath.
Decorate with the pastry trimmings.
To pack and freeze: Set on a double thickness of
kitchen foil placed over a baking sheet; chill until

Boeuf en croûte

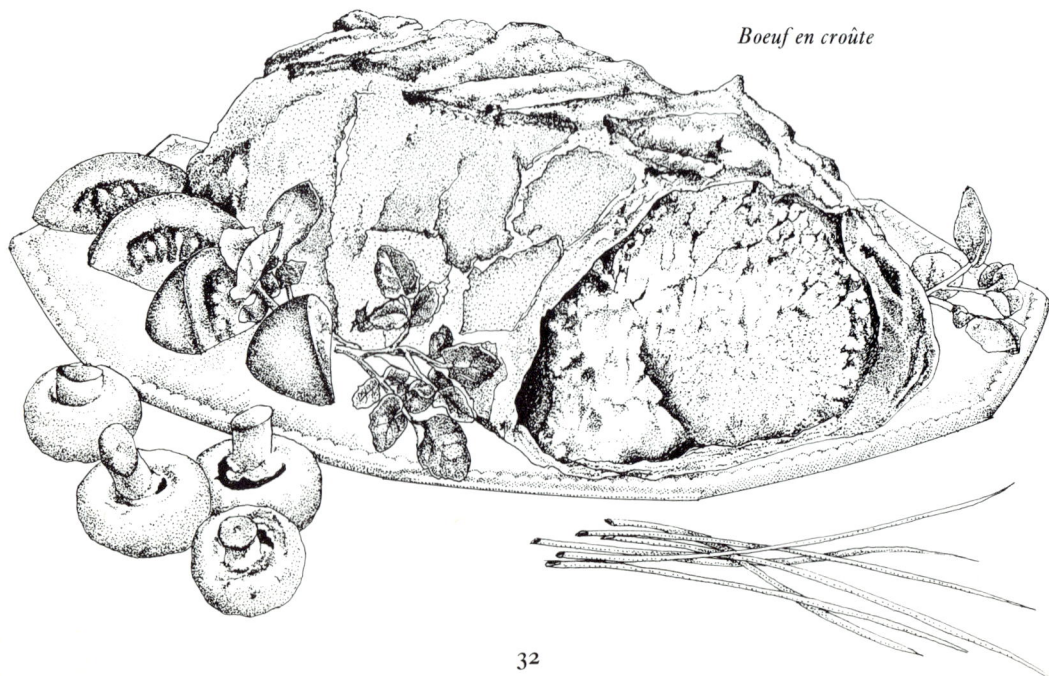

the pastry is firm, then wrap neatly in foil. Overwrap in a freezer bag, label and freeze until solid.

To thaw: Loosen seal, but leave wrapped in the refrigerator for about 36 hours.

To cook: Place on a wetted baking sheet in the oven, preheated to 450°F (mark 8) and bake for 15 minutes. Brush with beaten egg and continue to cook for a further 20–30 minutes. Garnish with sprigs of watercress. Slice at the table, using a sharp knife.

Blackcurrant sorbet

½ pt water
4 oz sugar
½ lb fresh or frozen blackcurrants
1 tsp lemon juice
2 egg whites

To make: Place the water in a saucepan with the sugar, bring to the boil and boil gently for 10 minutes. Cool. Meanwhile stew the blackcurrants in the minimum amount of water for about 10 minutes, then sieve and if necessary make the pulp up to ½ pt with water. Cool. Mix together the sugar syrup, lemon juice and fruit pulp. Pour into a freezer tray and freeze until nearly firm. Whisk the egg whites until stiff but not dry. Turn the frozen fruit mixture into a chilled bowl, break down with a fork and fold in the egg whites.

To pack and freeze: Turn the sorbet into a rigid plastic container or foil tray, cover, seal and return it to the freezer.

To serve: Soften the sorbet for a short time in the refrigerator – say, while serving the starter or main course of the meal. Scoop it out into glasses and serve alone, or combine it with a lemon sorbet.

Lemon sorbet

3 juicy lemons
8 oz caster sugar
1 pt water
1 egg white

To complete from freezer
mint leaf

Makes 3-portion packs for 3

To make: Pare the rind from the lemons free from all trace of white pith, and squeeze out the juice. Put the sugar, water and lemon rind in a pan. Dissolve the sugar over a low heat, bring to the boil and boil gently, uncovered, for 10 minutes. Leave to cool. Remove lemon rind then add the lemon juice and pour into one or more ice trays. Place in the freezer until mushy. Turn the mixture into a bowl, whisk the egg white until stiff but not dry and fold into the half-frozen lemon ice.

To pack and freeze: Turn into 3 × 1-pt rigid foil freezer trays or plastic containers. Cover with a lid and seal. Label and, if necessary, overwrap in a freezer bag. Freeze until solid.

To serve: Place in the refrigerator just before serving the starter or main course, so that the ice softens slightly. Decorate with a mint leaf, and serve half and half with blackcurrant sorbet. Two delicious varieties are (1) Serve it with Melba sauce (see page 15) or (2) Pour a little champagne round the water ice just before it is served.

FOR SIX

Home-made cream of vegetable soup with chives
French bread

Shoulder of lamb with apricot and ginger stuffing
*Roast potatoes   *Cauliflower sprigs

Crêpes suzette

Starred dishes are recipe suggestions or accompaniments
not suitable for freezing

Early evening 3 hours before guests arrive start cooking the joint of lamb. Take soup from freezer and turn it into a pan. Prepare vegetables. Remove cover from meat ½ hour before end of cooking time. Baste meat and potatoes (which should be added at the usual time). Make gravy. Heat the soup slowly; stir occasionally. Transfer crêpes suzette from the freezer container to a well-buttered ovenproof dish and dot with extra orange butter; cover with foil. Garnish the soup with chives before taking to the table. Cook vegetables. Set crêpes in the oven as the meat is removed; flambé them before serving.

Cream of vegetable soup

1 lb vegetables (see *Note*)
2 oz butter
½ pt white stock
salt, pepper
1 oz flour
1 pt milk

To complete from freezer
chopped chives

To make : Prepare the vegetables according to kind and roughly chop. Melt half the butter in a saucepan, add the vegetables and fry gently for 5 minutes without browning. Add the stock and simmer with the lid on the pan for 10–15 minutes, until the vegetables are tender. Season. Meanwhile, make a white sauce, using 1 oz butter, the flour and milk in the usual way. Pour the cooked vegetables and sauce into an electric blender and purée until smooth. Check seasoning.

Note Suitable vegetables are onions, leeks, tomatoes, broad beans, Brussels sprouts and green peas.

To pack and freeze : Allow to cool. Pour into a rigid polythene container, label and seal. Freeze rapidly until solid.

To use : Allow to thaw or reheat from frozen. Serve garnished with chopped chives.

Stuffed shoulder of lamb

3½-lb shoulder of lamb, boned

For the stuffing
4 oz fresh white breadcrumbs
2 oz shredded suet
4 pieces of stem ginger, finely chopped
**4 oz drained canned apricot halves,
    chopped, *or* 8 oz dried apricots, soaked
    overnight in water**
1 large egg, beaten
seasoning

Crêpes suzette

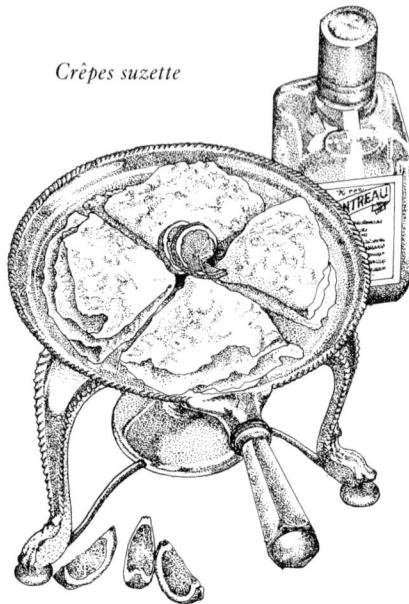

To make: Wipe the meat. Prepare the stuffing: mix the breadcrumbs and suet together, add the ginger and apricots, then bind with the egg; add a little seasoning. Stuff the lamb, using skewers to secure the joint into a good shape; tie with fine string, then withdraw the skewers.

To pack and freeze: Wrap in foil, making a neat parcel. Overwrap in a freezer bag, seal and label. Freeze rapidly until solid.

To serve: Cook uncovered from frozen. Seal the joint in hot fat in a preheated oven at 450°F (mark 8) for 20 minutes, reduce the heat, cover the meat and cook at 350°F (mark 4) for 1 hour per lb, basting frequently.

Crêpes suzette

For ¾ pt rich pancake batter
6 oz plain flour
pinch salt
2 small eggs
¾ pt milk
¾ oz butter, melted
butter or oil for frying

For filling
6 oz butter
6 oz caster sugar
grated rind of 2 large lemons
juice and grated rind of 2 large oranges
8 tbsps Cointreau

To complete from freezer
2–3 tbsps brandy

To make: Sift the flour and salt into a bowl. Make a well in the centre and break in the eggs. Add ½ pt milk and gradually incorporate the flour, beating. Add the remaining milk and the melted butter and beat until smooth and bubbly. Heat a little butter or oil in a 7-in thick based frying pan; pour off the excess and cook each pancake in the usual way, making 12 altogether. Transfer each one on to a square of freezer paper. Cream together the butter and sugar until light and fluffy. Gradually beat in the grated rinds, orange juice and Cointreau.

Note If at this stage the mixture separates, put the bowl over hot water and whisk rapidly, then place over cold water and whisk again, until smooth and well blended.

Spread half the filling across the pancakes; fold each in half and then in half again. Layer up the pancakes with freezer paper between them.

To pack and freeze: Wrap the pancakes in foil and then overwrap with freezer wrap, or place them in a rigid container. Seal and label. Pack the remainder of the orange butter in a small plastic container. Freeze until solid.

To use: Transfer the frozen pancakes to a shallow buttered ovenproof dish, dot with the reserved orange butter, then cover with foil. Heat through in the oven at 400°F (mark 6) for ½ hour. Warm 2–3 tbsps brandy in a small pan and ignite; pour it flaming over the pancakes and serve at once, decorated with a twist of orange.

FOR EIGHT

Danish cheese mousse

Stuffed loin of lamb

*Roast potatoes   *French beans

Ginger apple charlotte

Starred dishes are recipe suggestions or accompaniments not suitable for freezing

Countdown

Late afternoon Remove the mousse from the freezer and from its container. Allow to thaw at room temperature. Prepare vegetables. Begin to cook the lamb.

1 hour before the end of cooking place potatoes

around the meat and place Ginger apple charlotte to cook in the oven. Garnish the Danish cheese mousse. Cook beans and make the gravy.

Increase the heat to complete Ginger apple charlotte after serving the meat.

Danish cheese mousse

4 oz Danish blue cheese
$\frac{1}{4}$ pt double cream
$\frac{1}{2}$ oz shelled walnuts, chopped
$\frac{1}{4}$ oz powdered gelatine
2 tbsps water
salt, pepper and dry mustard
1 egg white
8 walnut halves

To complete from freezer
1 lettuce
fresh grapes or mandarin oranges

To make: Grate the cheese. Half whip the cream, add the nuts and fold in the cheese. Sprinkle the gelatine over the water in a saucepan, heat gently to dissolve it, then stir it into the cheese mixture; season to taste with salt, pepper and mustard. Stiffly whisk the egg white and fold into the mixture. Press it into a small container and garnish with walnut halves.

To pack and freeze: Wrap in foil, overwrap with freezer wrap, seal, label and freeze.

To use: Unwrap, remove from the container whilst still frozen. Cover lightly and allow the mousse to thaw at room temperature for 3 hours. Cut into 8 pieces. Serve on a bed of lettuce, garnished with fresh grapes or mandarin oranges.

Stuffed loin of lamb

3–3$\frac{1}{2}$ lb loin of lamb, boned
1 oz butter
1 small onion, skinned and finely chopped
4 rashers bacon, rinded and chopped
6 heaped tbsps fresh white breadcrumbs
grated rind and juice of 1 lemon
$\frac{1}{2}$ level tsp dried rosemary
1 small egg, beaten
salt and freshly ground black pepper

To complete from freezer
2–3 tbsps dripping for roasting
flour

Stuffed loin of lamb

For sauce
1 onion, skinned and finely sliced
1 level tbsp plain flour
$\frac{1}{2}$ pt stock
2 tbsps redcurrant jelly
lemon juice

To make: Wipe the meat and flatten it by lightly beating with a rolling pin or meat mallet. Melt the butter and gently sauté the onion and bacon. Mix the breadcrumbs, lemon rind and juice, rosemary and beaten egg. Season lightly with salt and pepper. Spread the stuffing over the meat, roll up and secure in several places with fine string.

To pack and freeze: Wrap tightly in foil, overwrap with freezer wrap, seal, label and freeze.

To use: Unwrap and cook from frozen. Melt the dripping in a roasting tin, dredge the meat lightly in flour and seal the joint in hot fat in a preheated oven at 450°F (mark 8) for 20 minutes. Reduce the oven to 350°F (mark 4), cover and cook for 1 hour per lb basting frequently.

Keep the joint warm on a serving plate. Pour off most of the fat from the roasting tin, leaving any sediment in the bottom. Add the sliced onion, and cook slowly, until just beginning to colour. Stir in the flour, stock and redcurrant jelly; sharpen the taste with a little lemon juice. Bring to the boil, stirring, cook for a few minutes and strain. Serve with the meat carved and arranged in a warm serving dish.

Ginger apple charlotte

2 lb cooking apples, peeled, cored and
    sliced
8 oz caster sugar
8 oz fresh white breadcrumbs
grated rind of 2 lemons
1 level tsp ground ginger
4 oz butter, melted

To complete from freezer
1 pt whipping cream or custard

To make : Blanch the apple slices in boiling water for 1–2 minutes, drain and cool. Mix together the sugar, breadcrumbs, lemon rind and ginger. Place alternate layers of crumbs and apples in two 1½-pt ovenproof dishes or rigid foil containers and after each layer pour over a little of the melted butter. End with a crumb layer.

To pack and freeze : Cover the dishes with foil, place in a freezer bag, seal, label and freeze.

To use : Unwrap the dishes, place on a baking sheet and cook from frozen in the oven at 350°F (mark 4) for 1 hour. Increase the oven temperature to 425°F (mark 7) and continue to cook for a further 15 minutes to brown the tops. Serve warm, with whipped cream or custard.

FOR EIGHT

Individual mixed fish pâtés   Hot toast fingers

Jugged hare with forcemeat balls and redcurrant jelly
*Boiled potatoes *Buttered green cabbage

Oranges with caramel

*Starred dishes are recipe suggestions or accompaniments
not suitable for freezing*

Countdown

The night before   Take jugged hare from the freezer, set on a plate, loosen the seal and leave in the refrigerator to thaw.

Early next morning   Take individual pâtés, oranges in syrup and forcemeat balls from freezer. Loosen seals and leave in the refrigerator to thaw.

Late afternoon   Prepare vegetables. Make the caramel and leave to set; crush it just before using as the finish to the dessert. Turn the oranges into a serving bowl or individual dishes.

Early evening   Put hare to reheat for 2–2½ hours in ovenproof dish. Place forcemeat balls in a baking dish and heat through in the oven for ½ hour before serving. Chop parsley for hare and potatoes. Garnish pâtés. Set potatoes to cook when guests arrive. Make toast at the last minute and wrap in napkin to keep warm. Put cabbage to cook while the first course is being eaten.

Mixed fish pâtés

7½-oz can salmon
7-oz can tuna
4 oz peeled prawns
6 oz fresh white breadcrumbs
4 oz butter, melted
juice and finely grated rind of 2 lemons
3 tbsps anchovy essence
½ pt single cream
salt and freshly ground black pepper

To complete from freezer
lemon twists

*Makes 8 individual pâtés and 1 large pâté for
4–6, for a future occasion*

To make : Remove any dark skin from the salmon and flake the flesh with its juices, together with the tuna and its oil. Chop the prawns roughly. Place the breadcrumbs in a bowl with the melted butter, lemon juice and rind. Add the flaked fish and prawns; stir in the anchovy essence, then the cream, and season to taste.

To pack and freeze : Spoon fish mixture into 8 ramekins or individual soufflé dishes (4-fl oz or 5-tbsps capacity). Place the remaining mixture in a 1-pt basin lined with foil. Place the dishes on a baking sheet and freeze rapidly until firm. Overwrap each individual dish in a freezer bag or freezer wrap; remove the large pâté from its basin, leave in the foil and overwrap. Seal, label and return the pâtés to the freezer.

To thaw : Allow 6 hours for small pâtés and 8–10 hours for the large one in the refrigerator, plus 1 hour at room temperature. In each case, unwrap the pâtes. Garnish with a twist of lemon.

Jugged hare

1 hare, cut into pieces, with the blood
$\frac{1}{4}$ lb bacon rashers, rinded
2 oz lard
1 onion, skinned and quartered
2 carrots, pared and sliced
2 sticks of celery, trimmed and chopped
2 pt stock
12 juniper berries
juice and thinly pared rind of 1 lemon
1 oz butter
1 oz plain flour
1 level tbsp redcurrant jelly
1 tbsp bottled Cumberland sauce
1 glass port or red wine

To make : Rinse and dry the hare joints. Keep the blood in a cool place until required. Cut the bacon into large pieces and place in a large frying pan with the lard. Add the joints of hare to pan and fry for about 10 minutes, turning frequently until sealed all over. Turn into a deep casserole with the onion, carrots and celery. Pour in the boiling stock and add the juniper berries, lemon juice and rind. Cover and cook in the centre of the oven at 325°F (mark 3) for about 3 hours. Strain the gravy from the casserole into a pan. Blend the softened butter and the flour to a paste, then stir, little by little, into the pan, to thicken the gravy.
Bring to the boil, stirring. Cool, then stir in the redcurrant jelly, Cumberland sauce and port or red wine. After 5 minutes, stir in the blood.

To pack and freeze : Arrange the joints of hare in a large ovenproof serving dish or casserole lined with foil. Pour the gravy over and leave to cool. Cover with double foil or place a piece of foil over the dish, replace the lid. Freeze rapidly until solid. Remove the foil package from the serving dish and place in a freezer bag. Seal and label.

To use : Unwrap and thaw in the covered casserole or serving dish at least 12 hours in the refrigerator. Reheat, covered, in the oven at 350°F (mark 4) for 2–2$\frac{1}{2}$ hours, till bubbling. Sprinkle the hare with chopped parsley. Place the hot forcemeat balls in the gravy or pile into a small casserole. Serve redcurrant jelly separately.

Forcemeat balls

3 rashers of streaky bacon, rinded and chopped
$\frac{1}{2}$ oz butter
4 oz fresh white breadcrumbs
2 oz shredded suet
2 tbsps chopped parsley
1 tbsp finely chopped onion
seasoning
beaten egg to bind
flour to coat

To make : Fry the bacon in the melted butter for 2–3 minutes. Add to the breadcrumbs, suet, parsley and onion in a bowl. Season, then bind with beaten egg. Form into 24 even-sized balls, using floured hands to shape them.

To pack and freeze : Arrange the forcemeat balls in a plastic box, with freezer paper between the layers to prevent them sticking. Freeze rapidly until firm. Alternatively, freeze them individually on a baking sheet, and when firm pack in a freezer bag and seal. Label, and store in the freezer.

To cook : Thaw and place in a well-greased baking dish; cook in the oven at 350°F (mark 4) for about 30 minutes. Turn them over occasionally.

Oranges with caramel

8 oranges
$\frac{3}{4}$ pt water
3 oz granulated sugar

To complete from freezer
8 level tbsps sugar
8 tbsps water
$\frac{1}{2}$ pt single cream

To make: Wipe oranges. Pare rind from 1 orange, free of any trace of white pith. Place water and sugar in a heavy-based pan, stir until the sugar is dissolved, then allow to boil. Add the orange rind and simmer for a further 5 minutes. Remove from heat and allow to become cold. Meanwhile peel the rest of the oranges, free of any white pith, and cut into thin slices, discarding pips.

To pack and freeze: Arrange the orange slices in a serving dish lined with foil or in a lidded foil container. Pour cold syrup over. Overwrap the serving dish with double foil and then put in a freezer bag, or cover with a lid if using a foil dish. Seal and label. Freeze until solid.

To thaw: Allow 12 hours in refrigerator or 6 hours at room temperature.

To serve: Make caramel by placing 8 level tbsps granulated or caster sugar and 8 tbsps water in a heavy-based pan. Over a low heat dissolve the sugar. Boil without stirring for about 7–10 minutes, until a pale caramel has formed. Pour on to a buttered or oiled baking sheet or marble slab,

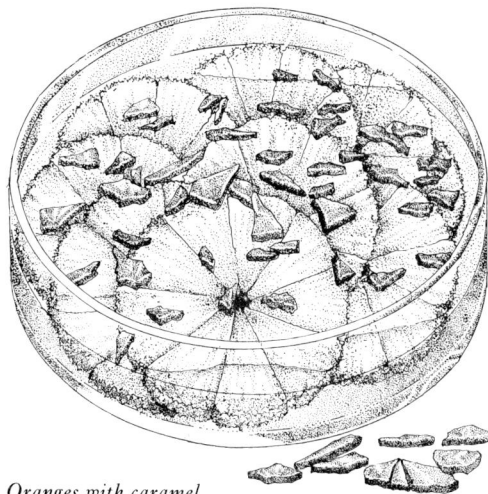

Oranges with caramel

leave till set, then break into small pieces. Scatter caramel over the oranges before serving with pouring cream.

FOR EIGHT

Chilled watercress soup   French bread

Boeuf stroganoff
*Boiled rice   *Sliced cucumber, tomato with chives
*Red and green peppers

Cheesecake topped with fruit

*Starred dishes are recipe suggestions or accompaniments
not suitable for freezing*

Countdown

The night before Remove cheesecake from freezer, unwrap, coat with crumb mixture, then loosely cover with foil. Leave in refrigerator. Remove the watercress soup, place in the refrigerator and allow to thaw overnight. Take stroganoff from freezer. Loosen cover on stroganoff, leave at room temperature.

Early evening Prepare all the salad ingredients and dressing (see page 83). Transfer stroganoff to a saucepan or flameproof dish ready to reheat. Invert cheesecake on to serving plate and decorate with fruit. Make apricot glaze and pour over cheesecake. Cook rice, drain in colander and leave over gently steaming water until required;

fork through occasionally. Leave the stroganoff to
heat through gently while the first course is being
eaten.

Chilled watercress soup

1 oz butter
1 small onion, skinned and chopped
2 bunches watercress
1½ pt chicken stock
1 pkt (2–3 servings) instant potato
1½ pt milk
seasoning

To complete from freezer
watercress leaves

To make : Melt the butter and sauté the onion for
2–3 minutes. Meanwhile trim away two-thirds of
the stem from the watercress. Wash, then dry and
chop. Add to the pan and stir well. Cover, then
cook over a gentle heat for a further 2–3 minutes.
Add the stock and simmer for 15 minutes. Bring to
the boil, remove from the heat, add the instant
potato and stir well before adding the milk. Return
to a gentle heat for a few minutes. Adjust
seasoning. Cool quickly.
To pack and freeze : Pour into a rigid plastic
container, label, seal and freeze. Alternatively,
pour into a freezer bag-lined container and when
frozen solid, remove the freezer bag from the
container, label and seal. Return to freezer.
To use : Defrost overnight in the refrigerator. (A
preformed package should be stood in a basin.)
Stir well, and just before serving, garnish with
watercress leaves. French bread or fried croûtons
are a good accompaniment.

Boeuf stroganoff

3 lb fillet of beef, thinly sliced
8 oz butter
2 medium-size onions, skinned and thinly
    sliced
½ lb mushrooms, sliced
2 oz flour
2 level tbsps tomato paste
2 × 15-fl oz cans consommé

To complete from freezer
salt and freshly ground black pepper
¼ pt soured cream

To make : Wipe the meat, then cut it into strips
approx 2 in by ¼ in. Melt 2 oz of the butter in a
large pan, fry the onions for 3–4 minutes, and lift
out on to a plate. Add a further 2 oz butter, reheat
and fry the meat a little at a time till sealed on all
sides; add 2 more oz butter during the cooking.
Remove the meat from the pan with a draining
spoon. Reheat the remaining pan juices and add
the remainder of the butter. Fry the mushrooms
for 2–3 minutes and add them to the meat and
onions. Stir the flour into the pan juices, followed
by the tomato paste. Continue to stir over a gentle
heat for 1 minute. Stir in the consommé, and cook
quickly.
To pack and freeze : Pack the meat, onion and
mushroom mixture separately from the sauce. Use
2 rigid foil containers for the meat, and 2 × 1-pt
freezer bags (preformed) for the sauce. Freeze
uncovered, until firm, then cover the foil containers with lids. Seal and label all the packets.
To thaw : Allow 16 hours in the refrigerator, 10
hours at room temperature.
To serve : Add the meat, etc, to the sauce, then
adjust the seasoning. Turn the mixture into a pan
or flameproof casserole to reheat on top of the
stove. When it is thoroughly heated through but
not boiling, turn it into the serving dish; keep
warm, and just before serving, stir in the ¼ pt
soured cream.

Cheesecake topped with fruit

3 oz butter
4 oz caster sugar
3 eggs, separated
2 oz ground almonds
1½ oz semolina
12 oz full-fat cream cheese
3 oz stoned raisins
grated rind and juice of 1 large lemon

To complete from freezer
8 oz digestive biscuits
4 oz butter, melted
fresh fruit or canned apricot halves
few shelled walnuts
apricot glaze

To make : Cream the butter and sugar; add the egg
yolks, almonds, semolina, cheese, raisins, lemon
rind and juice. Mix well. Fold in the stiffly whisked
egg whites.
Line the base of an 8½-in (3-pt capacity) cake tin

with waxed paper and pour in the cheesecake mixture. Place on a baking sheet and bake in the centre of the oven at 350°F (mark 4) for about 50 minutes, until set, reducing the heat to 300°F (mark 2) if it is browning too quickly.

Note This cake tends to sink on cooling.

To pack and freeze: Leave in the tin; when cold, cover with a lid or with kitchen foil. Overwrap, using a freezer bag, seal and label. Freeze until solid.

To thaw and serve: Unwrap the cheesecake, but leave it in its container. Allow 24 hours in the refrigerator, or 10 hours at room temperature. Meanwhile, crush the digestive biscuits in a bowl and mix in the melted butter. Spoon this mixture on to the top of the thawing cheesecake and smooth with a knife. Leave until the cheesecake is fully thawed and the biscuit-crumb top is set. Ease around the edges with a round-bladed knife. Invert the cake, crumb side down, on to a serving dish and decorate with fresh fruit in season, or with canned apricot halves, plus shelled walnuts. Top with apricot glaze (right).

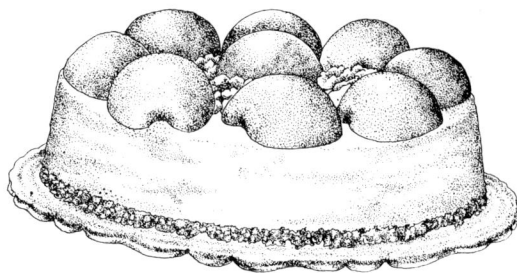

Cheesecake topped with fruit

Apricot glaze

Place ½ lb apricot jam and 2 tbsps water in a saucepan over a low heat and stir until the jam softens. Sieve the mixture, return it to the pan, bring to the boil, and boil gently until the glaze is of suitable consistency.

FOR EIGHT

Chicken liver pâté

*Grilled or fried mackerel

Gooseberry sauce and horseradish cream

*New potatoes   *Green salad

Danish peasant girl with veil

Starred dishes are recipe suggestions or accompaniments not suitable for freezing

Countdown

The day before   Order the fish, to collect on the day it is needed.

The night before   Take the sweet, pâté and gooseberry sauce from the freezer. Place in refrigerator and allow to thaw overnight.

Early evening   Flood the top of the pâté with melted butter and set in refrigerator, when set remove and leave at room temperature. Take the horseradish cream from the freezer, unwrap and leave at room temperature to thaw. Top the sweet with jam, whipped cream and chocolate; leave in the refrigerator or a cool place till required. Prepare the potatoes. Wipe and dry the fish. Wash the salad ingredients, drain and put them in a polythene bag or a bowl in the refrigerator until required. Measure ingredients for French dressing (see page 83) into a jar when required, shake well and pour over the salad. Transfer sauces for the fish to serving dishes. Set the potatoes to cook after the guests have arrived. Start cooking the mackerel on one side, turn them and leave to cook on second side whilst eating the pâté.

Chicken liver pâté

1½ lb chicken livers
1 medium-size onion
2 small cloves of garlic
3 oz butter
1 tbsp double cream
2 tbsps tomato paste
3 tbsps sherry or brandy
a little salt and freshly ground black
   pepper

To complete from freezer
2 oz butter, melted
sprig of parsley

Makes an 8-portion and a 2-portion pack

To make: Rinse the chicken livers in a colander. Dry on absorbent paper. Skin and finely chop the onion. Skin and crush the cloves of garlic. Fry the chicken livers in the hot butter until they change colour. Reduce the heat, then add the onion and garlic, cover and cook for 5 minutes. Remove from the heat and cool slightly. Add the double cream, tomato paste, sherry or brandy, and season to taste. Purée in an electric blender, or pass the mixture through a sieve to make a smooth texture.

To pack and freeze: Turn two-thirds of the

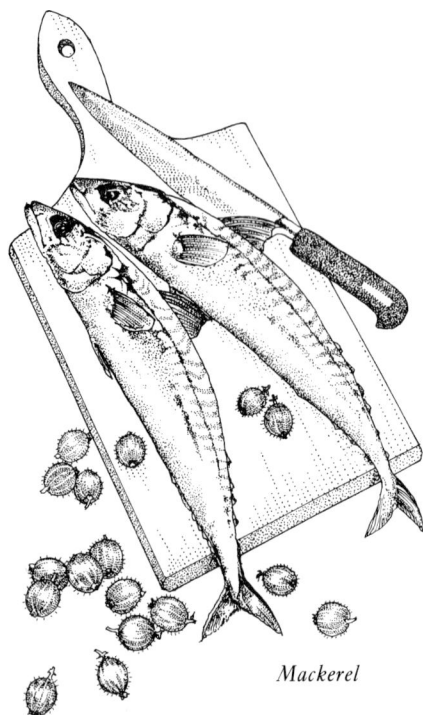

Mackerel

mixture into a ¾-pt soufflé dish or similar-sized rigid foil container, to give 8 portions. Package the remainder of the mixture for 2 portions. Allow to become cold before overwrapping or covering with a lid. Seal and label. Freeze until solid. Store for up to 1 month.

To thaw: Unwrap. Leave in the soufflé dish or transfer to a serving dish. Thaw overnight in the refrigerator and leave at room temperature for 1 hour before serving. When thawed, flood the top with melted butter and put in refrigerator to set.

To serve: Garnish with a sprig of parsley. Serve with Melba toast or crisp crackers.

Gooseberry sauce

2 lb gooseberries, topped and tailed
4 tbsps water
4 oz butter
sugar to taste

Makes 1½ pt

To make: Place the gooseberries in a colander and rinse well. Turn into a saucepan with the water and half the butter. Cover, and cook until the gooseberries are soft and pulpy. Pass them through a sieve or purée in an electric blender. Stir in the remaining 2 oz butter, cut into small pieces. Add a little sugar to taste if wished (though this can be added when the mixture is being reheated), but do keep the sauce on the sharp side. Allow to cool.

To pack and freeze: Divide into two ¾-pt portions (¾ pt is enough for 8 mackerel) and pack in rigid plastic containers or preformed freezer bags. Freeze until solid, seal and label.

To use: If serving hot, reheat from frozen in a heavy-based pan over a gentle heat, adding extra sugar if liked. To serve cold, allow to thaw in the refrigerator overnight, or at room temperature for 6 hours.

Horseradish cream: See page 28.

Danish peasant girl with veil

12 oz fresh breadcrumbs
6 oz soft light brown sugar
4½ oz butter
2 lb cooking apples, peeled, cored and
   sliced
juice of ½ lemon
sugar to taste

42

To complete from freezer
½ **pt double cream, whipped**
grated chocolate
raspberry jelly or strawberry jam, optional

Note The apples, lemon and sugar may be replaced by ½ pt prepared apple purée.
To make: Mix the breadcrumbs and sugar together. Fry in hot butter for 1 minute, then reduce the heat and cook, turning all the time, for about 10 minutes, till crisp and golden. Spoon on to absorbent kitchen paper. Meanwhile cook the apples in very little water with the lemon juice and sugar to taste, to give a thick pulp – about 10–15 minutes. Purée if wished. Cool thoroughly.
To pack and freeze: Put half the crumb mixture in a layer in a serving dish, then the purée and top with the remaining crumb mixture. For quicker thawing use a shallow rather than a deep dish.
To use: Allow to thaw overnight in refrigerator. Spread with jam (if used), cover with cream and sprinkle with grated chocolate.

Danish peasant girl with veil

Make a Cocktail Party Easy

Use a selection of our ideas, mixing them with some 'shop-bought' snacks

Sablés   Talmouse   Pâté fleurons
Anchovy twists   Cointreau dates   Maraschino roll-ups
Ham mustard pin wheels   Cheese cubes
Savoury choux buns

Sablés

4 oz plain flour
4 oz butter
4 oz mature Cheddar cheese, grated
pinch salt and dry mustard
freshly ground black pepper
beaten egg
a few chopped walnuts and almond halves

Makes about 60

To make : Sift the flour into a basin. Rub the butter into the flour with the fingertips until the mixture resembles fine crumbs, add the cheese and seasoning and work together to form a dough. Roll out the dough and cut into bite-size pieces. Brush with beaten egg and press the chopped nuts over the surface. Place the sablés on baking sheets and bake in the oven at 400°F (mark 6) for about 10 minutes. Cool on a wire rack.
To pack and freeze : Line a rigid polythene or foil container with freezer paper or waxed paper. Carefully pack the sablés, placing freezer paper between each layer. Cover with the lid or foil and overwrap with freezer wrap. Seal, label and freeze.
To use : Unwrap and refresh in the oven at 350°F (mark 4) for a few minutes.

Talmouse

7½-oz pkt 'boil in the bag' smoked haddock fillets
¼ pt frozen Béchamel sauce *(see page 83)*
salt and pepper
13-oz pkt frozen puff pastry, thawed
beaten egg

Makes about 12

To make : Cook the fish as directed on the packet. When cold, discard the skin and bones and flake the flesh. Heat the sauce gently in a saucepan until thawed and mix in the fish. Adjust the seasoning. Thinly roll out the puff pastry and stamp out as many rounds as possible, using a 3-in plain cutter, re-rolling as necessary. Brush the rim of each with beaten egg and place a little fish in the centre of each. Shape up the pastry into a tricorn. Seal the edges firmly, brush with more beaten egg, place on greased baking sheets and bake in the oven at 400°F (mark 6) for about 20 minutes until golden. Cool on a wire rack.
To pack and freeze : Line a rigid polythene or foil container with freezer paper or waxed paper. Carefully pack, placing freezer paper between the layers. Cover with the lid or foil and overwrap with freezer wrap. Seal, label and freeze.

To use: Unwrap and refresh in the oven at 350°F (mark 4) for 5 minutes.

Pâté fleurons

7½-oz pkt frozen puff pastry, thawed
beaten egg
4½-oz tube liver pâté
2 oz butter

Makes 50

To make: Roll out the pastry to about ⅛ in thick. Using a 1½-in fluted cutter, stamp out as many rounds as possible, re-rolling as necessary. Brush each round with beaten egg and fold over into a semi-circle. Place on baking sheets. Allow the fleurons to stand in a cool place for at least 30 minutes. Brush again with beaten egg before baking in the oven at 400°F (mark 6) for about 15 minutes. With a sharp knife, almost cut through the pastry to allow steam to escape. Leave to cool on a wire rack. Combine the pâté with the butter and beat well. Fill a piping bag fitted with a small star icing nozzle. Pipe in 'shell' shapes down centre of each fleuron (see illustration below right).
Alternatively, freeze the cooked fleuron cases, packing as for Talmouse; when required, unwrap them and refresh from frozen in the oven at 400°F (mark 6) for 5–8 minutes before filling with pâté.

Anchovy twists

scraps of puff pastry (about 3 oz)
2-oz can anchovy fillets
lemon juice
beaten egg

Makes about 25

To make: Roll out the pastry ⅛ in thick and cut into strips the length of each anchovy. Lay a strip of anchovy on each pastry strip and sprinkle with lemon juice. Twist the two together, brush with beaten egg, place on a baking sheet and bake in the oven at 450°F (mark 8) for 5–8 minutes. Cool on a wire rack.
To pack and freeze: Line a rigid polythene or foil container with freezer paper or waxed paper. Layer the anchovy twists, placing the freezer paper between each layer. Cover with the lid or foil and overwrap with freezer wrap. Seal, label and freeze.

To use: Unwrap the twists, allow to thaw for 1 hour or place in the oven still frozen and refresh at 350°F (mark 4) for about 5 minutes.

Cointreau dates

8 oz whole dates
6 oz soft cream cheese
grated rind 1 orange
2 tbsps Cointreau

To complete from freezer
grated rind of 1 orange

Makes about 20

To make: Split the dates, but do not cut them completely in half; remove the stones. Soften the cheese and blend it with the rind and Cointreau in a basin. Spoon or pipe a little of the mixture into each date.
To pack and freeze: Place the stuffed dates on a baking sheet, open freeze until firm. Wrap in a freezer bag, overwrap in another, seal and label. Return to the freezer.
To use: Unwrap and allow to thaw for 2–3 hours. Sprinkle with freshly grated orange rind. Serve in paper cases or on cocktail sticks.

A selection of cocktail snacks

Maraschino roll-ups

8-oz bottle Maraschino cherries
12–15 rashers streaky bacon, rinded

Makes about 30

To make : Drain the cherries. Cut each rasher of bacon in 2–3 pieces. Roll a piece of bacon round each cherry and secure each with a wooden cocktail stick.
To pack and freeze : Wrap in foil in a single layer, overwrap in freezer wrap, seal, label and freeze.
To use : Unwrap the roll-ups, place frozen under a preheated grill until crisp. Serve hot on cocktail sticks.

Ham mustard pin wheels

4 oz butter, softened
2 level tsps made mustard
4 oz cooked shoulder ham, sliced

Makes about 30

To make : Mix the butter and mustard together in a basin until well blended. Spread the mixture evenly on the ham slices and roll each up like a Swiss roll.
To pack and freeze : Pack the rolls in a rigid foil or polythene container. Cover with foil or a lid, overwrap with freezer wrap. Seal, label and freeze.
To use : Unwrap but keep loosely covered and thaw at room temperature for about 2 hours ; slice thinly and serve.

Cheese cubes

1-in cubes of white bread
beaten egg
a little milk
grated cheese

To make : Dip the bread cubes into the egg mixed with a little milk, then into the grated cheese. Place on a baking sheet.
To pack and freeze : Freeze unwrapped until firm.

Wrap in a freezer bag, seal and label. Return to the freezer.
To use : Unwrap the cubes and place on a baking sheet in the oven at 400°F (mark 6) for 15–20 minutes. Serve in a bowl, with chutney handed separately.

Savoury choux buns

1$\frac{1}{2}$ oz butter or margarine
$\frac{1}{4}$ pt water
2$\frac{1}{2}$ oz plain flour, sifted
2 eggs, beaten

To complete buns from freezer
4 oz cream cheese
2 oz softened butter
1 tsp lemon juice
Marmite
chopped parsley

Makes 24

To make : Place the butter and water in a saucepan, heat until the butter melts and then bring to the boil. Remove from the heat, tip in all the flour, beat well and cook gently until the mixture forms a ball and comes away from the sides of the pan. Cool slightly, then beat in the eggs a little at a time. Using a $\frac{1}{2}$-in plain vegetable nozzle, pipe out about 24 small walnut-sized balls of paste on to greased sheets. Bake in the oven at 400°F (mark 6) for 15–20 minutes until golden and crisp. Cool on a wire rack.
To pack and freeze : Spread on baking sheets and freeze until solid. Place in a freezer bag. Overwrap with another freezer bag. Seal, label and return to the freezer.
To use : Unwrap the buns and place on baking sheets. Refresh from frozen in the oven at 350°F (mark 4) for about 10 minutes. Cool on wire racks. For the filling, cream together the cheese, butter and lemon juice. Fill a forcing bag fitted with a small plain vegetable nozzle, make a hole in the base of each choux bun with the nozzle and pipe the filling into the buns. Glaze with a little Marmite and sprinkle with chopped parsley.

An Informal
'Dip and Spread'
Party

These recipes will cater for an evening drinks party for about 20 people.
The cool, creamy dips ask for a choice of dunkers – whole radishes, cauliflower florets,
carrot sticks, celery, spring onions – as well as pretzels, crisps and crackers. The rough
meaty texture of country-style pâté is an excellent contrast to the lighter dips.
Eat it – and the smooth liver pâté and taramasalata – with
coarse brown bread, pumpernickel, oatcakes
and crispbreads.

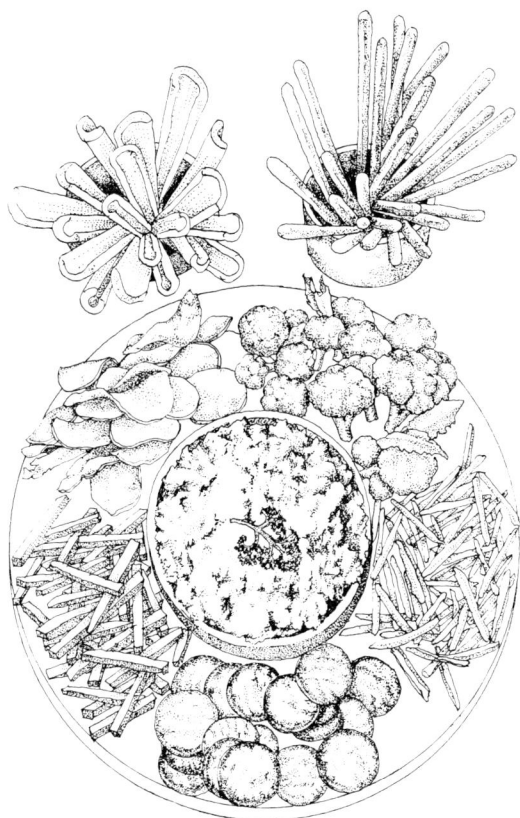

Country-style pâté

1 lb lean veal
1 lb lean belly of pork
$\frac{1}{2}$ lb lamb's liver
1 small onion, skinned and chopped
2 tbsps olive oil
2 tbsps wine vinegar
1 tbsp brandy or sherry
1 egg
salt and pepper
$\frac{1}{4}$ level tsp dried sage
6–8 rashers of streaky bacon, rinded and
    stretched

To complete from freezer
sprigs of parsley
2 tomatoes
lettuce leaves

To make: Cut the meats and liver into pieces and pass them through a coarse mincer. Add the onion, oil, vinegar and brandy to the meat mixture and blend well. Leave in a cool place for 2 hours. Add the beaten egg, seasoning and sage. Line an ovenproof 2-pt dish (such as a casserole, terrine or soufflé dish), with the rinded and stretched bacon rashers. Spoon in the meat mixture, smooth the top and cover with foil. Set in a roasting tin, with water to come half-way up the sides of the dish. Cook in the oven at 375°F (mark 5) for about 2$\frac{1}{2}$ hours. Remove from the water bath, and leave to cool with a weight on top. Pour off any excess liquid, if not jellied.
To pack and freeze: When cool, cover with foil, overwrap in freezer wrap, seal and label. Freeze rapidly until solid. Store for up to 1 month.
To thaw: Uncover, and loosely re-cover. Allow 16 hours in refrigerator plus 2 hours at room temperature.
To serve: Either serve whole, garnished with

parsley, or serve in slices, garnished with tomato wedges and/or crisp lettuce leaves.

Fresh salmon pâté

2 oz butter
2 oz flour
$\frac{3}{4}$ pt milk
1 bay leaf
salt and pepper
$\frac{1}{4}$ level tsp ground nutmeg
$\frac{1}{2}$ lb fresh haddock fillet, skinned
1 lb fresh salmon, skinned and boned
grated rind and juice of 1 lemon
1 tbsp chopped parsley
2 eggs, beaten
melted butter

To complete from freezer
slices of lemon
parsley

To make: Melt the butter in a pan, remove from the heat and stir in the flour. Cook for 2–3 minutes. Slowly add the milk, beating after each addition. Add the bay leaf, salt, pepper and nutmeg and boil gently for 2–3 minutes. Discard the bay leaf. Finely chop or mince the haddock and $\frac{3}{4}$ lb of the salmon; add to the sauce. Stir in the lemon rind and juice, parsley and eggs. Divide the mixture between 6–8 buttered individual soufflé dishes (3-fl oz). Brush the tops with melted butter. Slice the remainder of the salmon and decorate the tops. Place the dishes in a large roasting tin, pour in enough water to come half-way up the soufflé dishes and cook in the oven at 300°F (mark 2) for about 40 minutes. Leave to get cold.
To pack and freeze: Leave in the containers; wrap in foil, overwrap in freezer wrap. Seal and label. Freeze rapidly until solid. Store for up to 1 month.

To use: Thaw overnight in the refrigerator. Garnish with parsley sprigs and lemon slices.

Kipper pâté

12-oz pkt frozen kipper fillets in butter
2 oz butter
juice of $\frac{1}{2}$ lemon
4 tbsps double cream, lightly whipped
freshly ground black pepper

To complete from freezer
lemon wedges
sprigs of parsley

To make: Cook the kippers as directed on the packet. Remove from packet, discard any dark skin, then flake fish and juices with a fork; allow to cool. Melt butter and beat it into the fish, with lemon juice. (Use an electric blender for this if you have one.) Add cream and season with pepper.
To pack and freeze: Use a foil-lined or unlined 1-pt serving dish or similar-sized foil container. Spread the pâté evenly in the container. Freeze until firm, remove the pâté from the serving dish and overwrap; cover a foil container or unlined serving dish with a lid, or overwrap. Label and return it to the freezer.
To thaw: Leave loosely wrapped in the refrigerator for 12 hours or at room temperature for 6 hours.
To serve: Garnish with wedges or twists of lemon and sprigs of parsley.

Guacamole

3 small ripe avocados
3 tbsps fresh lemon juice
8-oz pkt full fat soft cheese
dash of Tabasco sauce
pinch cayenne pepper

To complete from freezer
paprika pepper
few olives to garnish

To make: Cut each avocado in half and remove the stones. Scoop out the flesh into an electric blender goblet, add the lemon juice and cheese and blend until smooth. Add a little Tabasco and cayenne pepper and beat again.
To pack and freeze: Turn the mixture into a foil container. Cover with foil and with freezer wrap, seal and label. Freeze rapidly until solid.
To thaw: Leave overnight in the refrigerator or at room temperature for 4–6 hours.
To serve: Dust with paprika pepper and garnish with olives.

Frikadeller

1 lb lean pork, finely minced
2 small onions, skinned and minced
1 level tsp salt
freshly ground black pepper
4 level tbsps flour
2 eggs, beaten
a little milk if necessary
fat for deep frying

Makes approximately 60

To make: Place the minced pork in a bowl. Add the onions, salt, pepper and flour. Bind with beaten egg and sufficient milk to give a soft but manageable mixture. Roll into small balls the size of a walnut. Heat the fat until it will brown a cube of bread in 1 minute. Cook the meatballs until brown – allow about 6 minutes. Drain on absorbent kitchen paper and cool.
To pack and freeze: Spread the frikadeller on baking sheets and open freeze rapidly until solid. Place in a freezer bag, overwrap by placing inside another freezer bag. Seal and label.
To reheat: Unwrap and place the frikadeller on a greased baking sheet. Reheat from frozen in the oven at 400°F (mark 6) for 15–20 minutes.
To serve: Use to dip into home made tomato or barbecue sauce (see page 83).

Ceviche

$1\frac{1}{4}$ lb filleted mackerel
juice of 2 lemons
1 large tomato, skinned and quartered
1 small onion, skinned and sliced
1 small green chili (or $\frac{1}{4}$ level tsp chili powder)
4 tbsps olive oil

To complete from freezer
few olives

To make: Wipe and skin the fish and remove any bones; place the fish in a shallow dish. Pour the

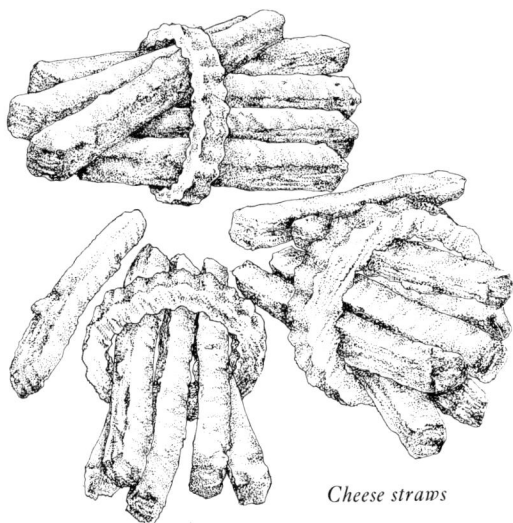

Cheese straws

salmon will cover 4 small slices of bread, and each can then be cut into 4 fingers. Squeeze a little lemon juice on each and dust with black pepper.

Cheese straws

4 oz plain flour
salt and cayenne pepper
2 oz butter
2 oz mature Cheddar cheese, grated
1 egg yolk
cold water to mix

To complete from freezer
1 egg white
grated Parmesan cheese
paprika pepper

Makes 40

To make: Season the flour with salt and cayenne and rub in the butter to give the texture of fine breadcrumbs. Mix in the cheese and egg yolk, with enough cold water to give a stiff dough. Roll out the pastry thinly and trim into oblongs 8 in long and $2\frac{1}{2}$ in wide. Put on to a greased baking sheet and cut into straws $2\frac{1}{2}$ in long and $\frac{1}{4}$ in wide, separating them as you cut. Roll out the remaining pastry and cut rounds with a 2-in plain cutter, then cut out the centre of the rounds with a $1\frac{1}{2}$-in plain cutter. Put all the shapes on to the baking sheet. Bake towards the top of the oven for 10–15 minutes at 400°F (mark 6) until pale golden in colour – watch carefully. Remove from the oven and cool slightly on the baking sheet before cooling completely on a wire rack.
To pack and freeze: Line a plastic box or foil container with freezer paper or waxed paper. Lift the straws into box and cover each layer with a piece of freezer paper. Cover with lid. Label, seal and freeze.
To reheat: Return the frozen straws to a baking sheet. Whisk 1 egg white until frothy, then use to brush the straws lightly. Sprinkle with Parmesan cheese and dust with paprika pepper, if liked. Refresh in the oven at 450°F (mark 8) for 10 minutes, until crisp and brown. Cool, then put a few straws into each ring before serving.

lemon juice over, cover and leave for 24 hours in the refrigerator, basting occasionally. Place the fish and juices in an electric blender with the tomato and onion. Cut a slice from the stalk end of the chili, discard the seeds and cut the chili into small pieces. Place in the blender, add the oil and blend until smooth.
To pack and freeze: Place in a foil container. Cover with foil, overwrap in a freezer bag. Seal and label. Freeze rapidly until solid.
To thaw: Leave in the refrigerator overnight, or at room temperature for about 8 hours.
To serve: Return the mixture to the blender, blend until smooth and season to taste. Turn into a serving dish and garnish with olives.

Smoked salmon

Allow 2 oz per head

To pack and freeze: Place the slices of smoked salmon between layers of freezer paper or waxed paper; make into a parcel, using foil. Place the parcel inside a freezer bag. Seal, label and freeze rapidly.
To thaw: Allow 8 hours in refrigerator or 4 hours at room temperature.
To serve: As a cocktail nibbler, serve on fingers of thinly sliced brown bread and butter; 4 oz smoked

Entertaining Buffet-style

Stuffed pancakes

*Dressed salad of red and green peppers, chicory and watercress

Pavlova with peaches

Cheeseboard and biscuits or French bread

*Starred dishes are recipe suggestions or accompaniments
not suitable for freezing*

Dressed salad

Allow 4 peppers, 6 heads of chicory and 4 bunches
of watercress for 12 people.
Seed and thinly slice the peppers, trim and slice the
chicory and thoroughly wash the watercress.
Place in large salad bowl and just before serving
toss with $\frac{1}{2}$ pt French dressing (see page 83).

Stuffed pancakes

1$\frac{1}{2}$ pt batter *(see page 35, batter for Crêpes suzette)*
lard for frying

For filling

12 oz onion, skinned and chopped
1$\frac{1}{2}$ oz butter
1$\frac{1}{2}$ lb minced beef
3 large carrots, grated
6 level tbsps cornflour
3 × 14-fl oz cans tomato juice
salt and freshly ground black pepper
**1$\frac{1}{2}$ level tsps each dried mixed herbs and
thyme**
3 tsps soy sauce

To complete from freezer
parsley

Serves 12

To make : Prepare 24 × 8-in pancakes in the usual
way. Sauté the onion in butter without browning.
Add the mince and cook until sealed, add the
carrots and cook for 5 minutes. Stir in the
cornflour, blended with the tomato juice, seasoning, herbs and soy sauce. Bring to the boil, reduce
the heat and simmer, covered, for 10 minutes.

Stuffed pancakes

When cold use to fill the pancakes and roll them up.
To pack and freeze: Place the pancakes in a foil dish, cover with foil, overwrap with freezer wrap. Seal, label and freeze.
To serve: Unwrap the pancakes. Place in a baking dish and cover loosely with foil; place in the oven at 375°F (mark 5) and heat for about 50 minutes.

Remove the foil cover and serve immediately, garnished with chopped parsley.

Pavlova See page 23. Make 2 for 12 people.

Cheeseboard Have a selection of cheeses, place on cheeseboard and serve with crisp celery or radishes. Place biscuits or bread in a basket.

INFORMAL BUFFET FOR 15

Sweet-sour pork

*Boiled rice   *Green salad

Mandarin treacle tart   Lemon cheese meringue

Starred dishes are recipe suggestions or accompaniments not suitable for freezing

Sweet-sour pork

4 lb lean pork
deep fat for frying
1 lb plain flour or equal quantities flour
  and cornflour
salt and pepper
8 eggs
6–8 tbsps milk

For the sauce
4 × 12-oz cans pineapple cubes
4 medium-size carrots, cut into matchstick
  pieces
2 green peppers, seeded and cut into strips
4 sticks of celery, trimmed and finely sliced
4 tbsps olive oil
3 oz cornflour
8 tsps soy sauce
2 oz brown sugar
6 tbsps vinegar

To complete sauce from freezer
2 small onions, skinned and finely sliced
cooking oil, to sauté onions
2 oz flaked almonds, toasted

Serves 15

To make: Trim the pork into neat chunky cubes, and set the fat to heat. Sift the flour (or flour and cornflour) into a bowl with the salt and pepper. Drop in the eggs, add the milk and beat to a smooth, thick batter. Coat the pork cubes in batter and fry for 15 minutes, or until the batter is golden and the pork cooked through. Set aside to cool quickly.

Meanwhile prepare the sauce. Drain and remove the juice from the cans of pineapple; make up to 1½ pt with water. Fry the carrots, pepper and celery in the hot oil for 2–3 minutes, stirring frequently. Blend the cornflour with a little of the pineapple liquid to a smooth paste, then stir in the remaining liquid. Add to the pan and bring to the boil stirring, until the sauce is smooth and clear. Add the soy sauce, brown sugar, vinegar and pineapple cubes. Allow to cool quickly.

To pack and freeze: Open freeze the pork cubes by spreading out on a baking sheet so that they are not touching, and freeze. When frozen place in a foil container. Pour the sauce into a party-size foil or polythene container. Cover with lids or a double layer of foil, overwrap with a freezer bag, seal and label. Freeze rapidly until solid. Or pour the sauce into a large freezer bag in a rigid container, freeze then remove the freezer bag, seal and label.

To cook: Thaw pork cubes overnight in the

refrigerator. Remove the wrappings and place the sauce in a flameproof casserole or heavy-based pan over gentle heat. Break the sauce down with a wooden spoon as it begins to thaw (keeping it covered between times). Add 2 skinned, finely sliced and sautéd small onions and 2 oz toasted almonds to the hot sauce just before serving. Set the pork balls on baking sheets in the oven at 425°F (mark 7) for about 15 minutes, until really hot.

To serve : Place the pork balls on a serving dish and pour some of the sauce over them. Surround with freshly boiled rice, with a further bowl for the remaining sauce.

Mandarin treacle tart

6 oz shortcrust pastry (ie made with 6 oz flour, etc)
3 oz fresh white breadcrumbs
8 oz golden syrup
grated rind and juice of 1 lemon
11-oz can mandarin oranges, drained

To complete from freezer
icing sugar

Make 3 to serve 15

To make : Roll out the pastry and use to line a 7-in fluted flan ring on a baking sheet. Combine the breadcrumbs, syrup, lemon rind and juice and mandarin oranges. Pour into the case. Bake in the oven at 400°F (mark 6) for 50–60 minutes. Remove the flan ring and cool on a wire rack.

To pack and freeze : Open freeze until firm then wrap in foil. Place in a freezer bag, seal, label and return to the freezer.

To use : Either Unwrap, place on a baking sheet and allow to stand at room temperature for 5–6 hours, then reheat in the oven at 350°F (mark 4) for about 30 minutes. *Or* Unwrap, place on a baking sheet and reheat from frozen in the oven at 375°F (mark 5) for about 1 hour. Serve warm, dredged with icing sugar.

Lemon cheese meringue

6 oz shortcrust pastry (ie made with 6 oz flour etc)

For the filling
2 oz butter
3 oz caster sugar
3 egg yolks
2 oz plain flour
8 oz cottage cheese, sieved
grated rind and juice of 2 lemons

To complete from freezer
3 frozen egg whites, thawed
2 oz caster sugar
1 level tbsp granulated sugar

Make 3 to serve 15

To make : Roll out the pastry and use it to line an 8-in flan ring on a baking sheet. Cream the butter and sugar until light and fluffy and beat in the egg

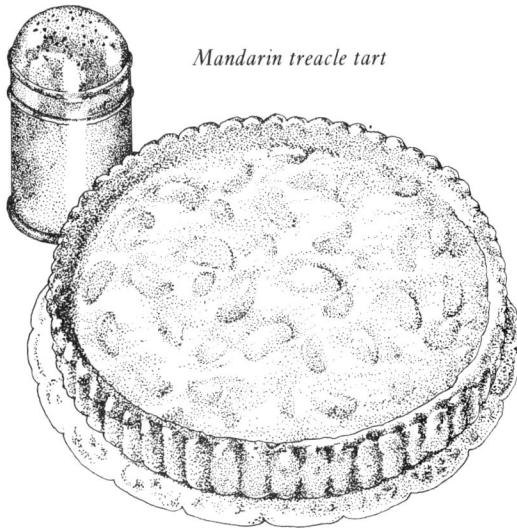

Mandarin treacle tart

53

yolks. Fold in the flour, cheese, lemon rind and juice. Turn into the pastry case.

To pack and freeze: Open freeze the flan with the ring in position until firm. Remove the ring and wrap the flan in foil and place it in a freezer bag. Seal, label and return it to the freezer.

To use: Unwrap the flan, place it on a baking sheet and replace the flan ring. Thaw for 2 hours at room temperature. Bake in the oven at 400°F (mark 6) for 35 minutes. Whisk the egg whites until stiff and fold in the caster sugar. Remove the flan from the oven; reduce the temperature to 300°F (mark 2). Pile the meringue on top of the flan, making sure it covers the filling right to the edge but does not overlap (if it overlaps it is difficult to remove the flan ring). Sprinkle the granulated sugar over the top and return it to the oven for a further 20–25 minutes until the topping is tinged with brown. Remove the flan ring and serve at once.

SUMMER BUFFET FOR 40

This delicious, rather grand menu will serve about 40 people in all. Quantities vary from recipe to recipe because we're not suggesting your guests should wade their way through everything – the idea is to choose. If you *know* all your friends love quiche though, you'd best make 2 or 3 and cut down on something else.

Cream of lemon soup with walnut sablés

Curried cucumber soup

Blue cheese quiche   Lemon-stuffed roast veal

Terrine of duck   Jellied chicken pie

*Tossed green salad   *Tomato and onion salad

Chocolate and rum cheesecake pie

Lemon cheesecake pie

Starred dishes are recipe suggestions or accompaniments not suitable for freezing

Cream of lemon soup

3 oz butter or margarine
1¼ lb onions, skinned and sliced
1¼ lb carrots, pared and sliced
6 pt turkey or chicken stock
6 lemons
2 bouquets garnis
2 oz arrowroot
salt
freshly ground black pepper

To complete from freezer
¾ pt single cream

Serves 20

To make: In a large saucepan melt the butter. Add the sliced vegetables. Cook gently until tender, stirring frequently. Pour over the stock, bring to the boil, reduce the heat and simmer. Using a vegetable peeler, thinly pare the rind from the lemons. Pour boiling water over the rind, leave for 1 minute then drain. Add the rind and juice of the lemons and bouquets garnis to the pan contents. Cover and cook for 1 hour or until the vegetables are really soft. Remove the bouquets garnis. Purée the soup a little at a time in an electric blender. In a clean pan, blend the arrowroot with a little of the soup then add the remainder, stirring. Season and bring to the boil stirring. Cool rapidly.

To pack and freeze: Freeze in preformed freezer

bags or rigid containers – about 1 pt quantities for quick thawing.

To use: Reheat slowly from frozen; when thawed bring to the boil, reduce the heat, check seasoning and add the cream. Reheat but do not boil.

Walnut sablés

8 oz plain flour
8 oz butter
8 oz mature Cheddar cheese, grated
pinch each salt and dry mustard
beaten egg
a few chopped walnuts

Makes about 60

To make: Sift the flour into a basin. Rub the butter into the flour with the fingertips until the mixture resembles fine crumbs. Add the cheese and seasonings and work them together to form a dough. Roll out the dough on a lightly floured board. Neaten the edges and cut into 2-in squares. Cut the squares in half diagonally and brush with beaten egg; sprinkle with nuts and press them lightly into the pastry. Place the sablés on baking sheets and bake at 400°F (mark 6) for about 10 minutes. Cool on a wire rack.

To pack and freeze: Pack into freezer bags or a rigid container, seal, label and freeze.

To use: Refresh in the oven for a few minutes, then cool on a wire rack.

Curried cucumber soup

5 oz butter
1½ lb onions, skinned and very finely chopped
5 cucumbers
3 level tsps curry powder
5 level tbsps plain flour
8 chicken stock cubes and 6 pt water *or* 6 pt home made chicken stock
salt and pepper

To complete from freezer
3 pt creamy milk

Serves 20

To make: In a large pan, such as a deep preserving pan, melt the butter, add the onion and fry without colouring for 10 minutes. Add the cucumber,

coarsely grating it straight into the pan, and fry for a further 5 minutes. Add the curry powder and flour, cook a few minutes. Add stock, stir and bring to the boil. Adjust seasoning. Cover, reduce heat and simmer for 30 minutes. Cool rapidly.

To pack and freeze: Freeze in preformed freezer bags or rigid containers – about 1 pt quantities for quick thawing.

To use: Reheat slowly from frozen, adding the milk. Adjust seasoning.

Blue cheese quiche

8 oz plain flour
salt and freshly ground black pepper
2 level tsps icing sugar
5 oz butter at room temperature
3 standard eggs and 1 egg yolk
4 oz onion, skinned and chopped
3 oz lean streaky bacon
6 oz Stilton cheese
¼ pt double cream
¼ pt milk

To complete from freezer
chopped chives *or* parsley *or* spring onions

Serves 10

Blue cheese quiche

To make : Sift the flour, salt and icing sugar into a bowl. Add the butter, which should be soft enough for your fingers to leave an impression on it if you press it. Add one egg and with the fingers of one hand gradually work the butter and egg together, then draw in the flour to make a firm but pliable paste, working it in lightly with the tips of your fingers. Place in a polythene bag and chill for 30 minutes.

Fry the onion and rinded, chopped bacon until the fat runs and the onion is tender. Cool. Bring the pastry back to rolling temperature and roll out thinly to line a 9-in loose bottomed French fluted flan ring. Prick the base and bake blind at 325°F (mark 3) for about 30 minutes until the base is dry. Scatter the onion mixture over the base, crumble the cheese over. Lightly whisk the cream, milk, remaining eggs and yolk. Season carefully as the cheese may be salty. Pour in enough custard mixture to nearly fill the case. Return the quiche to the oven at 375°F (mark 5) and cook for about 35 minutes. Cool in the turned off oven for about 15 minutes. Chill.

To pack and freeze : Carefully remove the quiche from the flan ring (or if preferred leave in the tin), wrap in foil and overwrap in a freezer bag or use freezer wrap. Seal, label and freeze.

To use : Unwrap and return the quiche to the flan ring. Thaw at room temperature for about 3 hours and then refresh at 350°F (mark 4) for about 30 minutes. Serve warm (not hot) or cold garnished with chopped chives, parsley or spring onions.

Lemon-stuffed roast veal

3 lb breast of veal, boned weight
salt, milled pepper
3 tbsps lemon juice
1 oz fresh white breadcrumbs, toasted
1 level tsp dried rosemary
1 level tbsp chopped parsley
1 clove garlic, skinned and crushed
1 oz butter
1 egg, beaten
4 thin slices lemon
4 rashers fat streaky bacon, rinded
1 tbsp cooking oil

Serves 10

To make : Flatten out the meat and season with salt, pepper and 1 tablespoon lemon juice. In a bowl combine the breadcrumbs, rosemary, parsley, garlic, melted butter, ¼ level teaspoon salt, black pepper and egg. Using a palette knife, fill the pocket in the meat with stuffing and spread any remaining stuffing over the meat. With scissors remove the rind from the lemon slices and arrange down the centre of the meat. Roll up and secure with string at intervals. Lay the bacon across the meat. Place the oil and the rest of the lemon juice in the base of a casserole just large enough to take the joint. Place the meat on top, cover and roast in the oven at 350°F (mark 4) for 3½ hours – until tender. Baste occasionally. Cool quickly.

To pack and freeze : Slice the veal and rearrange the slices together; wrap in foil, overwrap with a freezer bag, seal, label and freeze.

To use : Thaw overnight in the refrigerator, unwrapped.

Terrine of duck

4–4½ lb oven-ready duck, roasted
1 lb belly of pork
1 lb pie veal
¼ lb back pork fat
1 clove garlic, skinned and crushed
½ level tbsp salt
freshly ground black pepper

To complete from freezer
1 small orange, sliced
aspic jelly

Serves 10

To make : Let the duck go cold, discard the skin and carve off every scrap of flesh. Trim the surplus fat from the belly pork and trim the veal. Cut 4–6 thin strips from the pork fat and reserve for garnish. Put the duck, pork, veal and the rest of the pork fat through the mincer twice. Add the garlic, salt and pepper and mix well. Put this mixture in a 2–2½-pt terrine or casserole. Arrange the pork fat strips in a lattice on top, then cover. Stand the terrine in 1 in water in a roasting tin and cook at 350°F (mark 3) for 2 hours. Remove from the oven, take off the lid and leave for 15 minutes, cover with folded foil and press with weights. Chill.

To pack and freeze : Remove the weights, place the covered dish in a freezer bag, seal, label and freeze.

To use : Thaw in the refrigerator for a day, garnish with slices of fresh orange and a layer of aspic jelly. Leave at room temperature for 1 hour before serving.

Jellied chicken pie

2 × 3-lb fresh oven-ready chickens
$\frac{1}{2}$ lb lean streaky bacon
$\frac{1}{2}$ lb onions, skinned and thinly sliced
$\frac{1}{2}$ lb mushrooms, stalked and sliced
2 tbsps chopped parsley
$\frac{1}{2}$ level tsp mixed herbs
salt and pepper
4 tbsps water
12 oz rich shortcrust pastry (12 oz flour,
    7 oz butter, etc)
beaten egg to glaze

Serves 10

To make : Skin the chickens, carve off all the flesh and cut into pieces. Rind the bacon and cut it into small pieces. Layer the chicken, bacon, onions and mushrooms in a shallow ovenproof dish about 12-in by 9-in by 2-in. Sprinkle the layers with parsley, mixed herbs and seasoning. Place a pie funnel in the centre. Add the water. Roll out the pastry and use to make a lid; decorate with the trimmings made into leaves.
To pack and freeze : Chill to set the pastry. Pack the pie carefully in a freezer bag or use freezer wrap. Seal, label and freeze.
To use : Thaw in the refrigerator for about 7 hours. Brush the pastry with beaten egg. Cut a slit in the pastry lid, place the pie on a baking sheet and cook in the oven at 350°F (mark 4) for about $1\frac{1}{2}$ hours. Cool quickly and chill.

Chocolate and rum cheesecake pie

For biscuit base
6 oz sweetened digestive biscuits
3 oz caster sugar
3 oz butter, melted

For the filling
4 oz cottage cheese, sieved
4 oz cream cheese
4 oz caster sugar
4 oz plain chocolate
$\frac{1}{2}$ oz powdered gelatine
8 fl oz water
2 tbsps rum
$\frac{1}{4}$ pt double cream

To complete from freezer
$\frac{1}{4}$ pt double cream
1 tbsp milk or whipping cream
chocolate curls
icing sugar

Serves 8

To make : Crush the biscuits in a basin with the end of a rolling pin. Add 3 oz sugar, and mix in the butter. Press round the base and sides of an 8-in flan dish with the back of a metal spoon. Chill until firm. Cream the cheeses together and add the remaining sugar. Melt the chocolate in a basin over hot water. Dissolve the gelatine in the water in a cup over a saucepan of boiling water. Add the chocolate and gelatine to the cheese and beat well. Chill until almost set. Add the rum. Lightly whip the cream and fold it into the chocolate mixture. Pour the filling into the prepared biscuit pie case. Chill until firm.
To pack and freeze : Cover with freezer wrap, label and freeze.
To use : Leave to thaw, covered, at room temperature for 6–8 hours. Lightly whip $\frac{1}{4}$ pt double cream and 1 tablespoonful milk or whipping cream, swirl this over the chocolate filling and decorate with curls of chocolate dusted with icing sugar.

Lemon cheesecake pie

6 oz digestive biscuits
3 oz caster sugar
3 oz butter, melted
$1\frac{1}{2}$ × 1-pt pkts lemon jelly
3 tbsps water
2 eggs, separated
$\frac{1}{4}$ pt milk
grated rind of 2 lemons
4 tbsps lemon juice
12 oz cottage cheese
$\frac{1}{2}$ oz caster sugar
$\frac{1}{4}$ pt double cream, whipped

For decoration
$\frac{1}{4}$ pt double or whipping cream, optional

Serves 8

To make : Crush the digestive biscuits, place them in a bowl and combine with the sugar and butter. Use to line a 9-in (2-pt) shallow open pie plate.

Press the crumbs into place with the back of a spoon. Chill. In a small pan over a low heat, dissolve the jelly in the water. Don't boil. Beat together the egg yolks and milk, pour on the jelly, stirring, and return the mixture to the saucepan. Heat for a few minutes without boiling. Remove from heat and add the lemon rind and juice. Cool until beginning to set. Stir in the sieved cottage cheese, or blend the jelly mixture and unsieved cheese in an electric blender. Whisk the egg whites stiffly, add the sugar and whisk again until stiff.

Fold quickly into the cheese mixture, followed by the $\frac{1}{4}$ pt whipped cream. Spoon into the crumb crust, piling up slightly. Chill until set, then top with whipped cream before freezing (or freeze without the cream and add the topping when serving).

To pack and freeze: Open freeze, wrap in foil or freezer film, overwrap with freezer wrap, seal and return to freezer.

To use: Leave to thaw, covered, at room temperature for 6–8 hours. Decorate.

FORMAL BUFFET FOR 35–40

This is just the menu for a small wedding, or some
such occasion. You'll find the food makes a most impressive,
colourful spread – particularly if you use
white linen tablecloths.

Baked salmon trout

Smoked trout and/or mackerel with mustard sauce

Boeuf en croûte with horseradish cream

Boned turkey with apricot and ginger stuffing

*Coleslaw

*Celery and potato mayonnaise

*Tomato and onion salad   *Californian salad

French bread and butter

Pavlova   Charlotte russe

Summer fruit salad with cream

*Starred dishes are recipe suggestions or accompaniments
not suitable for freezing*

Baked salmon trout

a 4–5 lb salmon trout, cleaned but whole

Serves 6–8

To pack and freeze: Coat the fish with layers of ice (see page 18 as for Trout meunière), wrap closely in kitchen foil, overwrap with a freezer bag, seal and label. Freeze until solid.

To thaw: Allow 24–30 hours in the refrigerator. *To cook:* Unwrap, then wrap loosely in a sheet of well-buttered foil, adding 1–2 sprigs of parsley and 1–2 lemon slices, with a bay leaf if desired. Set on a baking sheet, and cook in the oven at 300°F (mark 2), allowing 30–35 minutes per lb, or until the flesh is just beginning to show signs of coming away from the bone. Allow to cool in the foil.

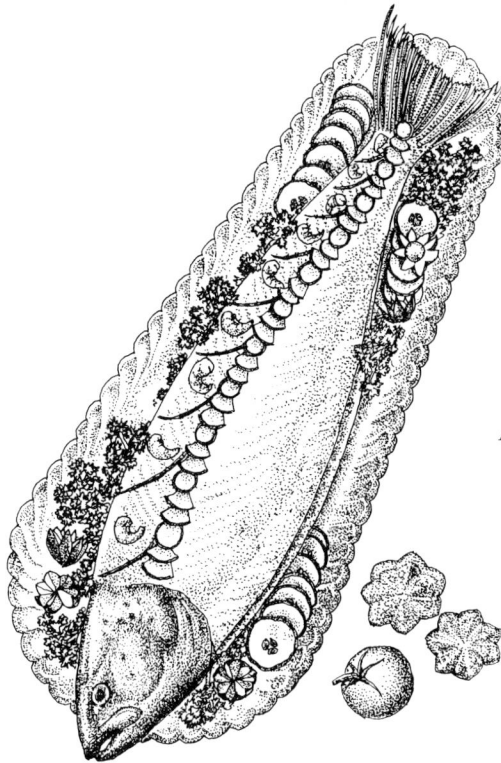

Baked salmon trout

To coat in aspic and decorate
¾ pt aspic jelly
radishes, cucumber, parsley, stuffed olives,
    tomato skin, shrimps, etc, to garnish

Remove the skin from the body of the fish, leaving
on the head and tail. Make up the aspic jelly
according to the packet instructions and, when it is
just beginning to thicken, coat the fish thinly.
Decorate the fish, using thin rounds of radish,
strips of cucumber skin or thin cucumber slices,
diamonds or strips of tomato skin, rounds of olive,
sprigs of parsley, picked shrimps, etc, all dipped in
aspic. Cover the decorated fish with further layers
of aspic until the garnish is held in place. Leave to
set. Serve with a mixed salad and mayonnaise, or as
desired.
Note The easiest way to carry out the glazing
process is with the fish on a wire rack and with a
large plate underneath to catch the drips. When
the surplus aspic is set, chop it on damp
greaseproof paper with a sharp knife, and use as an
additional garnish.

Smoked trout and mackerel (see page 28);

allow 6 of each, and serve with the sauces suggested
for accompanying them.

Boeuf en croûte (see page 32); prepare double
quantities. Cook the beef on the day and allow to
become cold. Serve in slices, garnished with plenty
of fresh watercress and accompanied by horseradish cream (see page 28).

Boned and stuffed turkey

a 10-lb fresh oven-ready turkey, cleaned
    and boned – approx 6 lb meat
apricot and ginger stuffing *(see page 34, but
    use treble quantities)*
4 oz softened butter
seasoning

Serves 25–30

Ask your poulterer to clean and bone out the
turkey for you. Stuff with the prepared mixture
and tie or sew into shape.
To cook: Weigh, then put in a roasting tin and
cover with 4 oz softened butter and some

59

seasoning. Cover with foil and cook in the oven at 325°F (mark 3), allowing 35–40 minutes per lb. Half an hour before the end of the cooking time remove the foil, baste well, pour off the excess fat, then increase the oven temperature to 425°F (mark 7), so that the skin of the bird will brown. Lift on to a plate and allow to cool quickly.

To pack and freeze : Wrap in a double thickness of kitchen foil, overwrap with a freezer bag, seal and label. Freeze rapidly until solid.

To thaw : Allow about 2 days in refrigerator.

To serve : Cut in neat slices; the platter may be garnished with apricots, if you wish.

Coleslaw

This recipe is a menu suggestion not suitable for freezing

1 hard white cabbage, washed and finely shredded
3 onions, skinned and finely chopped
4 level tbsps chopped parsley
1 pt salad cream (approx)
1 level tbsp sugar
salt and pepper
a few drops of vinegar or lemon juice

Combine the cabbage, onion and parsley in a large bowl. Mix the salad cream with the sugar, salt and pepper and add enough vinegar or lemon juice to sharpen the flavour. Toss with the salad in the bowl until lightly coated, adding a little more salad cream if necessary.

Celery and potato mayonnaise

This recipe is a menu suggestion not suitable for freezing

4 lb potatoes, peeled
2 large onions, skinned and finely chopped
head of celery, trimmed and finely chopped
1 pt mayonnaise
4 level tbsps chopped parsley

Boil the potatoes in salted water until just cooked. Drain well and cut into ½-in dice. Mix the diced potato, onions and celery together. Add the mayonnaise and mix well. Sprinkle with chopped parsley before serving.

Tomato and onion salad

This recipe is a menu suggestion not suitable for freezing

4 onions, skinned and sliced
12 firm tomatoes, skinned and sliced
⅓ pt French dressing *(see page 83)*
chopped chives

Arrange the onions and tomatoes alternately in a shallow dish. Pour the dressing over and serve sprinkled with the chives.

Californian salad

12 tomatoes, thinly sliced
½ cucumber, diced
6 sticks celery, finely chopped
salt and pepper
6 bananas, thinly sliced
6 red-skinned apples, cored and diced
juice of 1½ lemons
salad cream or mayonnaise to bind
lettuce or watercress

Sprinkle the tomatoes, cucumber and celery with salt and pepper. Soak the bananas and apples in the lemon juice for about 10 minutes and drain. Mix all the ingredients together and bind lightly with the salad cream or mayonnaise. Serve on a bed of lettuce or surrounded with watercress.

Charlotte russe

¼ pkt lemon jelly
⅛ pt boiling water
1½ pkts sponge fingers
½ pt double cream, whipped
1 pt cold custard (made with 2 level tbsps custard powder and 1 level tbsp caster sugar, with 1 pt milk)
2–4 tbsps Kirsch or Cointreau, optional
1 level tbsp gelatine
3 tbsps water

To complete from freezer
remaining ¾ pkt lemon jelly
¼ pt double or whipping cream, whipped

Serves 8: for this buffet make 2

To make : Make up the jelly with the boiling water and set 4 tbsps of it in the bottom of a straight-sided mould ($2\frac{1}{2}$–3 pt capacity). Allow to set. If necessary, trim the sides of the sponge fingers, then brush the edges with the remaining liquid jelly. Line the sides of the mould with the fingers, pressing them closely together. Support the fingers with a ring of foil. Combine the whipped cream, custard and liqueur (if used). Dissolve the gelatine in the water in the usual way and add to the cream mixture. When this is on the point of setting, pour it into the centre of the mould, lifting the foil away, and allow to set, then trim off any surplus from the top of the sponge fingers.

As a variation, you can make a fruit-flavoured cream for the centre – use $\frac{1}{2}$ pt fruit purée to replace $\frac{1}{4}$ pt of the custard.

To pack and freeze : Carefully wrap the whole dish in kitchen foil. When quite frozen, overwrap with a freezer bag, seal and label.

To thaw : Unwrap, then stand the mould in very hot water for a minute or two to loosen the base. Cover with a plate and invert to unmould the charlotte, but leave the mould over it to give it support. Leave it in the fridge for 14–16 hours to thaw, or leave it at room temperature for approximately 8 hours. Meanwhile make up the remaining $\frac{3}{4}$ pkt of lemon jelly and leave it in a shallow dish to set.

To serve : Remove the mould. Chop the jelly and spoon two-thirds of it round the base of the sweet. Pipe extra whipped cream round the top edge and fill the centre with the remaining chopped jelly. Tie a ribbon round before serving.

Pavlova See page 23; make two.

Summer fruit salad See page 66; make twice the amount given.

Teenage Party Menus

ITALIAN PIZZA PARTY FOR 12

This is the sort of fresh, exciting menu most teenagers
love and it's also straightforward, you will enjoy making it, too.
Quantities will feed about 12 hungry people,
with a choice of pudding.

Iced cream of tomato soup garnished with chives

*Quick pizza   *Watercress and chicory salad*

Iced zabaglione   Lemon sorbet (see page 33)

*Starred dishes are recipe suggestions or accompaniments
not suitable for freezing*

Iced cream of tomato soup

8 oz onion, skinned
6 cloves
2 × 2-lb 3-oz cans tomatoes
sprig parsley
2 bay leaves
2 level tsps salt
freshly ground black pepper
½ level tsp freshly grated nutmeg
4 oz butter
6 level tbsps flour
1½ pt milk
½ pt light stock

To complete from freezer
6 tbsps single cream
chopped chives

To make: Cut the onion into small chunks. Stud

one piece with cloves. Place the onions, tomatoes
with their juice, the sprig of parsley, bay leaves,
seasoning and nutmeg in a saucepan. Bring to the
boil, reduce the heat, cover and simmer for 1 hour.
Melt the butter in another pan and blend in the
flour. Cook the roux for 2–3 minutes before
gradually stirring in the milk, to give a smooth
sauce. Bring to the boil, stirring, reduce the heat
and simmer for 5 minutes. Remove the bay leaves,
cloves and parsley from the tomato mixture and
purée it in an electric blender. Pass the purée
through a fine sieve to remove the tomato pips.
Add it to the white sauce. Blend them well
together. Stir in stock, check seasoning and cool.
To pack and freeze: Pour into a rigid polythene
container allowing 1 in headspace. Cover with a lid
or foil. Freeze until solid. Overwrap with
freezer wrap. Seal, label and return to the freezer.
To use: Allow to thaw overnight in the refrigerator. Just before serving stir in the cream and
sprinkle with chives.

Watercress and chicory salad

This recipe is a menu suggestion not suitable for freezing

4 bunches watercress
6 heads chicory
½ pt French dressing *(see page 83)*

Wash and drain the watercress. Trim and finely slice the chicory. Toss well together. Pour over the French dressing just before serving.

Quick pizza

1 lb 8 oz self raising flour
2 level tsps salt
½ pt cooking oil
12–16 tbsps water
4 onions, skinned and chopped
4 × 14-oz cans tomatoes, drained
8 level tsps mixed dried herbs
4 oz butter
1 lb Cheddar cheese, grated
40 olives (black or stuffed green)
4 × 2-oz cans anchovies

Makes 16

To make : Mix the flour and salt and stir in 4 tbsps of the oil and enough water to mix to a fairly soft dough. Divide into 16 pieces. Roll out each into 4-in rounds and fry on both sides in the remaining oil in large frying pans (or see variations, below). Drain well. Make the topping by frying the onion, tomatoes and herbs in the butter. Spread a quarter of the grated cheese over the dough base, with the tomato mixture. Sprinkle the remaining cheese over. Slice the olives, wash and drain the anchovies, and arrange in a lattice design on top.

To pack and freeze : Place each pizza on a large square of foil. Freeze rapidly, uncovered, until they are firm. Once they are solid, wrap foil around the pizzas, secure and label. Overwrap in a freezer bag.

To serve : Remove the bag, loosen the wrappings and leave lightly covered with foil. Bake from frozen in a preheated oven at 350°F (mark 4) for about 45 minutes, until the cheese has thoroughly melted; if necessary grill to brown the top for 1–2 minutes.

Variations

(i) Roll out the pizza dough into an oblong to fit a well oiled baking sheet measuring 10 in by 8 in. Bake in the oven at 450°F (mark 8) for 15 minutes

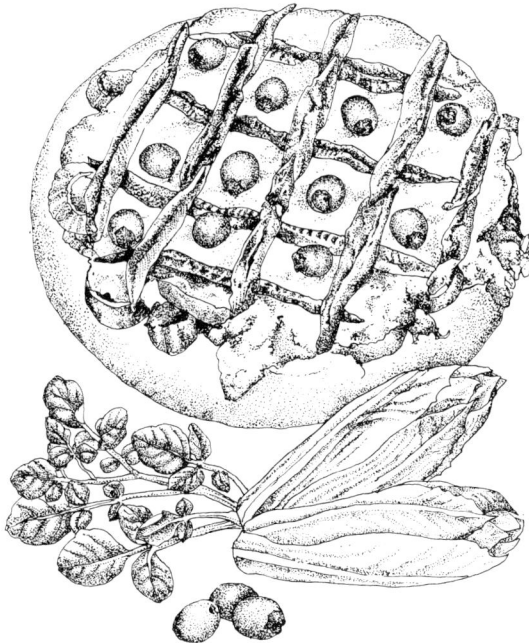

Quick pizza

until well risen and golden brown. Cover with the tomato and onion topping, then arrange overlapping slices of salami (2 oz) to cover the topping. Cut 4 oz cooked ham into neat strips and use to make a lattice effect over the salami. Freeze and wrap as basic recipe.

To cook: Unwrap. Cover loosely with foil. Bake from frozen in a preheated oven at 350°F (mark 4) for about 1 hour. Arrange half slices of tomato in between the lattice to complete the decoration and sprinkle with a little grated cheese, if liked. Put under a hot grill for 1–2 minutes before serving. (ii) Make up 4-in rounds as in basic recipe. Cook and top with tomato mixture, arrange 2 sardines on each round and garnish with olives. Freeze and wrap, reheat and serve as basic recipe.

Iced zabaglione

4 egg yolks
4 oz caster sugar
6 fl oz Marsala

To make: Beat the egg yolks to a pale cream and mix in the sugar and Marsala. Cook in a double saucepan until the custard coats the back of the spoon. Pour the custard into 6 individual ramekin dishes and cool.

To pack and freeze: Freeze until firm. Wrap each dish in foil and overwrap all in a freezer bag.

To serve: Remove the wrappings and serve straight from the freezer, accompanied by crisp wafer or boudoir biscuits.

GREEK PARTY FOR 12

Taramasalata   Freshly made toast
*Moussaka   *Cucumber and mint salad*
Home made ice cream (see page 67) with ratafias
**Fresh fruit*

Starred dishes are recipe suggestions or accompaniments not suitable for freezing

Taramasalata

12 oz fresh smoked cod's roe, skinned
⅓ pt olive oil
juice of 1½ lemons and a little grated rind
2 level tsps grated onion
2 level tbsps chopped parsley
freshly ground black pepper

To make: Place the roe in a bowl with half the olive oil and leave for 10 minutes. Pass it through a fine sieve, or blend in an electric blender until smooth, gradually adding the lemon juice and the remaining oil. Turn the mixture into a bowl, then stir in the lemon rind, onion, parsley and freshly ground black pepper.

To pack and freeze: Turn into a 1-pt foil dish. Cover with foil, overwrap with freezer wrap. Seal, label and freeze.

To thaw: Allow to thaw in the refrigerator for 8

hours or overnight, or for about 4½ hours at room temperature.

To serve: Unwrap and turn into a serving dish. Sprinkle with more parsley, or add twists of lemon. Eat with hot toast or crackers.

Moussaka

4 lb aubergines, trimmed and thinly sliced
salt
½ pt olive oil
4 oz butter
10 medium-size onions, skinned and thinly sliced
4 lb minced raw lean lamb
2 × 2¼-oz cans tomato paste
seasoning
2 × 15-oz cans tomatoes
2 bay leaves

64

Taramasalata

To complete from freezer
2 pt cheese sauce *(see Béchamel sauce recipe, page 83)*
grated Parmesan cheese
chopped parsley

To make: Spread the aubergines out on a large plate or tray, sprinkle with salt and leave for at least 1 hour. Pour off the liquid which collects and dry all the slices with absorbent kitchen paper. Fry the aubergines in several lots in the oil for about 10 minutes, turning them frequently. Meanwhile melt the butter in another pan and sauté the onions until soft. Place the minced lamb in a bowl and stir in the tomato paste and seasoning. Pass the tomatoes with their juice through a sieve, or purée in an electric blender. Line two large, shallow ovenproof casseroles with foil. Arrange in them layers of aubergine, lamb and onion, adding the bay leaves and finishing with aubergine. Pour half of the puréed tomatoes over each. (There should be enough room left for the cheese sauce to be poured on afterwards.) Cover, and cook in the oven at 350°F (mark 4) for 1 hour. Cool quickly.
To pack and freeze: Freeze the moussaka in the casseroles, then take out by easing the foil away from the dishes; cover with foil, overwrap in a freezer bag. Seal and label and freeze.
To use: Return to original casseroles and cook from frozen. Set the dishes on baking sheets and cook in the oven at 375°F (mark 5) for $1\frac{1}{2}$ hours. Meanwhile prepare or thaw the cheese sauce. Pour the sauce over the moussaka and sprinkle with a little Parmesan cheese. Return the moussaka to the oven and cook uncovered for a further $\frac{1}{2}$–1 hour, until the sauce is golden. Sprinkle with chopped parsley.

Cucumber and mint salad

This recipe is a menu suggestion not suitable for freezing

2 medium-size cucumbers
salt, pepper
3 tbsps chopped mint
$\frac{1}{2}$ pt natural yoghurt

Finely dice the cucumber. Place in a serving bowl and season lightly.
Add the chopped mint and yoghurt. Mix them well together and chill for 1 hour before serving.

Chili con carne   Boiled rice
Cucumber salad
Summer fruit salad with cream
Home made ice cream and Melba sauce (see page 15)

Chili con carne

6 lb minced beef
6 tbsps cooking oil
4 large onions, skinned and chopped
4 green peppers, seeded and chopped
4 × 15-oz cans tomatoes
salt and pepper
6 level tsps chili powder
2 tbsps vinegar
2 level tbsps sugar
2 × 2¼-oz cans tomato paste

To complete from freezer
2 × 15¼-oz cans red kidney beans

To make: Fry the beef in the hot oil until lightly browned; add the onion and peppers, then fry for a further 5 minutes, until soft. Stir in the tomatoes, seasoning, chili powder blended with the vinegar, sugar and tomato paste. Cover and simmer for 1 hour. Cool.

To pack and freeze: Spoon into 2 large foil dishes. Cover with foil. Overwrap with freezer wrap. Cover, label and seal. Freeze until solid.

To use: Cook from frozen in the oven at 400°F (mark 6) for about 1¼ hours. When thawed, stir in the kidney beans.

To serve: Serve with freshly boiled rice. Allow 2 oz rice per person.

Summer fruit salad

1 lb raspberries
1 lb strawberries
4 oz granulated sugar
1 pt water
2 oranges

Serves 6 (Note As there is a choice of pudding we have given quantities for half the party)

To make: Pick over the fruit, removing any stalks or hulls. Rinse only if it is felt necessary. Drain thoroughly keeping the different fruits separate. Dissolve the sugar in the water over gentle heat, stirring all the time. Finely pare the rind from the oranges, free of white pith, and add to the syrup; boil rapidly for 5 minutes then allow to become completely cold. Remove the orange rind and add the squeezed orange juice.

To pack and freeze: Pack the strawberries and raspberries into separate rigid foil containers, allowing 1 in headspace in each. Pour syrup over both and place the containers on a baking sheet. Freeze uncovered until firm, then cover with a lid, seal and label.

To thaw: Allow 18 hours in the refrigerator or 8 hours at room temperature:
Combine the raspberries and strawberries in syrup ½ hour before serving.

To serve: Spoon into glass dishes and serve with thick cream.

Chili con carne

Home made ice cream

½ pt milk
1 vanilla pod
1 whole egg
2 egg yolks
3 oz caster sugar
½ pt double or whipping cream

Makes about 1 pt (*Note* As there is a choice of pudding we have given quantities for half the party)

To make: Bring the milk almost to the boil and infuse the vanilla pod for about 15 minutes; remove the pod. Cream the whole egg, yolks and sugar together until pale. Stir in the flavoured milk and strain the mixture through a sieve back into the saucepan. Heat the custard very slowly, stirring all the time, until it just coats the back of a spoon. Pour the cold custard into a shallow foil container and freeze until mushy. Remove from the container and beat well. Fold the lightly whipped cream through it. Continue freezing until mushy; beat it a second time, and freeze again until firm.

To pack and freeze: Cover with foil, overwrap with freezer wrap. Seal, label and return to the freezer.

To use: Remove from the freezer. Place in refrigerator for 6–8 hours, to 'come to' before serving. Spoon Melba sauce over before serving (see page 15 for sauce recipe).

Variations

Chocolate Add 2 oz plain chocolate and (optional) ½ oz unsweetened chocolate, dissolved in the milk, to give a stronger flavour.

Praline Add 2 oz crushed praline, nut brittle or toasted hazelnuts at the end of the second beating.

Coffee Add 2 tsps coffee essence to the mixture when cold.

Pineapple Drain a 15-oz can of pineapple titbits. Purée the pineapple in an electric blender, or crush it well. Beat it into the half-frozen ice cream.

Hot Fork Suppers

CURRY PARTY FOR 12

Quantities given serve 12 people unless otherwise stated, for example, where there is a choice of 2 or more dishes.

Chicken, beef and prawn curries
*Boiled rice
*Side dishes
Summer fruit salad (see page 66)
Lemon sorbet (see page 33)

Starred dishes are recipe suggestions or accompaniments not suitable for freezing

Basic curry sauce

3 Spanish onions
3 cooking apples
6 oz butter
2 tbsps cooking oil
4 oz curry powder
3 level tsps curry paste
3 oz plain flour or cornflour
3 pt stock
4 tbsps sweet chutney
3 level tbsps tomato paste
juice of 1 lemon

Makes 4 pt

To make: Skin and chop the onions and apples. Heat the butter and oil in a pan, add the onions and apples and cook gently without browning for 5–8 minutes. Stir in the curry powder and paste and cook for 5 minutes, stirring occasionally, to bring out the full flavour. Stir in the flour or cornflour and cook for 1 minute, stirring all the time, before adding the stock. Bring to the boil, still stirring. Add the chutney, tomato paste and lemon juice, cover, and cook gently for about $\frac{3}{4}$ hour. Cool quickly.
To pack and freeze: Put in heavy freezer bags in plastic containers to shape. 1 pt quantities are

practical. Freeze, then overwrap with foil or place in another polythene bag. Seal and label.
To thaw: Thaw the sauce in a double saucepan or in a heavy-based saucepan over gentle heat. Break up, using a wooden spoon, as the sauce thaws. Check the seasoning.
Use to complete the following recipes.

Chicken curry

4-lb oven-ready fresh chicken (6 joints)
1–2 oz seasoned flour
2 oz butter
2 tbsps cooking oil
1 oz cornflour
2 pt unseasoned chicken stock
2 level tsps tomato paste
salt and pepper

Serves 4–6

To make: Joint or remove flesh of chicken. Wipe and toss in seasoned flour. Fry in the hot butter and oil in a large frying pan until brown all over – about 10 minutes. Lift them into a large casserole. Blend the cornflour to a smooth paste with some of the stock, then add the remaining stock and pour into the juices in the pan; stir until the sauce thickens. Add the tomato paste to improve the colour of the sauce, and stir in a little seasoning. Pour the sauce over the chicken, cover and cook in the oven at 350°F (mark 4) for $\frac{3}{4}$ hour. Cool quickly.

To pack and freeze: Line a roasting tin with foil and pour in the mixture. Freeze until solid, then overwrap with foil or place in a large freezer bag. Seal and label. Replace in the freezer.

To thaw and serve: Allow 24 hours in the refrigerator. Add to 1 pt thawed curry sauce in a large casserole. Taste for seasoning, cover and heat through in the oven at 375°F (mark 5) for $1\frac{1}{2}$ hours, until bubbling. Pour off the curry sauce into a saucepan and boil to reduce it by half. Arrange the chicken in a serving dish and spoon the reduced sauce over it.

Beef curry

$1\frac{1}{2}$ lb stewing steak
1 oz seasoned flour
2 oz butter
$\frac{1}{2}$ pt beef stock
1 level tsp tomato paste
seasoning

Serves 4–6

To make: Wipe the beef, trim and cut it into neat pieces. Toss in seasoned flour, then fry in melted butter until brown on all sides. Stir in the stock and tomato paste. Allow to come to the boil, then pour into a casserole. Cover and cook in the oven at 325°F (mark 3) for $1\frac{1}{2}$ hours, until fork tender. Cool as quickly as possible.

To pack and freeze: Turn into foil dishes, or freeze in the casserole. Overwrap or cover with a lid. Label and freeze.

To thaw: Allow 14 hours in the refrigerator. Add to 2 pt thawed curry sauce and taste for seasoning.

To cook: Heat gently in a saucepan or place in covered casserole in the oven at 375°F (mark 5) for 1 hour, until bubbling.

Prawn curry

For each 2–3 servings, add 4–6 oz peeled prawns to 1 pt thawed curry sauce. Heat through gently in a saucepan, or in a covered casserole in the oven at 375°F (mark 5) for $\frac{1}{2}$ hour, or until really hot.

Boiled rice

Allow 2 oz long grain rice per head. Cook in boiling, salted water for about 15 minutes. Drain, and rinse with boiling water. When dry and fluffy, turn it on to a serving dish and sprinkle with paprika pepper.

Beef curry and side dishes

Side dishes

Chop 3 large eating apples, sprinkle with lemon juice and dust with paprika pepper.
Cut 4 bananas in slices, sprinkle with lemon juice and dust with paprika pepper.
Peel and dice a cucumber. Mix with 2 × 5-fl oz cartons of plain yoghurt.
Bowl of 4–6 oz salted peanuts.
Bowl of mango chutney.
3 peeled and chopped tomatoes with 1 tbsp finely chopped onion and a little parsley.
Freshly made poppadoms or chapatis.

PAELLA PARTY FOR 12

Gazpacho

*Paella   *Dressed green salad*

French bread

Pineapple creams

Starred dishes are recipe suggestions or accompaniments not suitable for freezing

Gazpacho

3 lb tomatoes, skinned, quartered and seeded
3 onions, skinned and quartered
1½ cucumbers, chopped
12 tbsps wine vinegar
6 tbsps olive oil
6 tbsps red wine

To complete from freezer
2 cloves garlic, skinned and crushed
2 × 15-oz cans tomato juice
2 level tbsps chopped parsley

To make: Purée the tomatoes, onions and cucumbers in an electric blender. Pour the mixture into a bowl and stir in the vinegar, oil and red wine.
To pack and freeze: Pour the gazpacho into a rigid polythene container allowing 1 in headspace. Cover with a lid or foil. Seal, label and freeze until solid.
To thaw: Leave at room temperature for about 6 hours.
To serve: Skin and crush the garlic, add to the gazpacho together with the tomato juice; adjust seasoning and add the parsley.

Paella

4-lb oven-ready fresh chicken
3½ pt stock from the chicken
1 lb onions
3–4 tbsps cooking oil
3 oz butter
2 lb American long grain rice
2 pkt saffron powder
2 bay leaves
salt and freshly ground black pepper
7-oz can sweet red peppers, drained and chopped

To complete from freezer
2 cloves garlic, skinned and crushed
2 oz butter
8 oz peeled prawns
8 oz pkt frozen peas
4 oz pkt frozen sliced beans
10-oz jar mussels, drained
1 level tbsp chopped parsley
few tbsps white wine
lemon wedges
black olives
whole prawns

70

To make : Cook the chicken in 4 pt water with a few flavouring vegetables for about 1–1½ hours (or until tender). Lift it from the saucepan and allow to cool on a plate, then strip off the flesh and cut it into small pieces. Keep the stock. Skin, slice and chop the onions and fry them in the hot oil and butter for 4 minutes. Add the rice and cook over a gentle heat for 2–3 minutes, stirring so that the rice absorbs the fat. Add the saffron powder, then half the stock, the bay leaves and some seasoning. Allow to bubble over a gentle heat for about 10 minutes or until tender but not soft, adding more stock as required to keep the rice from drying out. Stir in the chicken pieces. Remove from the heat and allow to become cold.

To pack and freeze : Spoon into party size foil containers or foil-lined dishes. Divide the red peppers between the dishes, laying the pieces decoratively on top. Cover with the lids or foil, seal and label. Overwrap with freezer bags. Freeze rapidly until solid.

To thaw : Loosen the wrappings and leave overnight in the refrigerator.

To cook : Fry the crushed garlic in the butter for 4 minutes. Line 1 or 2 large roasting tins with foil and divide the melted butter and garlic between them. Stir in the thawed rice and chicken mixture. Add the peeled prawns. Cover loosely and put in the oven at 375°F (mark 5) for 15 minutes. Meanwhile cook the frozen peas and sliced frozen beans in boiling water for 2 minutes only; drain well. Add both to the paella in the tins, with the mussels, and parsley. Heat for a further 15 minutes. Add a few tbsps white wine if it seems dry.

To serve : Garnish with lemon wedges, black olives and whole prawns. Serve at once, with a green salad and French bread.

Pineapple creams   See page 31; make double quantity.

Paella

PASTA PARTY FOR 12

Quantities serve 12 people unless otherwise stated, for example, where there is a choice of 2 dishes or more.

Chicken tettrazini   Lasagne   Macaroni cheese
*Dressed green salad   *Tomato and onion salad
French bread and butter
*Fresh pineapple slices
*Fresh fruit – peaches or apricots, or as available
Liqueur cherry brûlée

*Starred dishes are recipe suggestions or accompaniments
not suitable for freezing*

Chicken tettrazini

1½ lb cooked chicken flesh (from a 6-lb
   boiling fowl)
12 oz spaghetti
½ lb button mushrooms
3 oz butter
15-oz can tomatoes, drained, optional

For the sauce
3 oz butter
3 oz plain flour
1 pt chicken stock
1 pt milk
seasoning
¼ pt white wine
4 oz grated cheese

Serves 6

To make : Cut the chicken meat into pieces. Cook
the spaghetti according to the packet directions
and drain thoroughly. Pour boiling water over it to
keep the strands separate. Rinse and drain the
mushrooms and sauté them in the melted butter
for 5 minutes. Prepare the sauce as follows. Melt
the butter in a pan, stir in the flour and cook for 1
minute. Remove from the heat and gradually stir
in the stock and milk. Bring to the boil, stirring
continuously, and cook for 2 minutes. Add
seasoning and the wine, then cool slightly before
stirring in the cheese. Add half the sauce to the
chicken pieces and the remainder to the drained
cooked spaghetti, with the mushrooms. Leave
until cold.

To pack and freeze : Line a large shallow ovenproof
dish with foil; spoon in the spaghetti mixture. Top
with tomatoes (if used), then cover with chicken in
sauce. Freeze until solid; remove from dish and
wrap the foil over. Place in a large freezer bag,
seal and label. Return it to the freezer.

To cook : Unwrap, then reheat in the serving dish,
with a light foil covering; allow about 1½ hours in
the oven at 350°F (mark 4). Fork through a few
times during the reheating. Garnish with chopped
parsley.

Lasagne

**Bolognese sauce, made with 1 lb minced
   beef** *(see page 84)*
4 oz lasagne verdi
1 tsp oil
¼ level tsp salt
½ lb cottage cheese
2 large eggs, beaten

To complete from freezer
1 pt cheese sauce
grated Parmesan cheese

Serves 6

To make : Have the Bolognese sauce ready. Put the
lasagne in a large pan half-filled with boiling water
to which the oil and salt have been added; cook for
12 minutes, or until *al dente*. Drain in a colander
and pour plenty of cold water on it. Dry on
absorbent kitchen paper. Place the cottage cheese
in a bowl with the beaten eggs.

To pack and freeze : Line a large shallow ovenproof
dish with foil and oil this. Line the base with half
the lasagne. Cover with all the Bolognese sauce.
Top with the remaining lasagne and the cottage
cheese and egg mixture. Freeze rapidly until set,
remove from the dish, overwrap with foil, seal and
label. Store in the freezer.

To cook : Remove the foil cover and replace the
lasagne in the original ovenproof dish. Top with

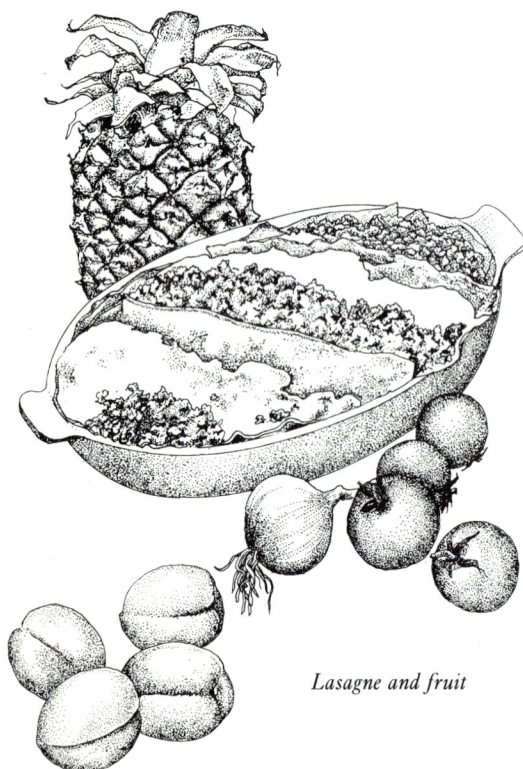

Lasagne and fruit

1 pt well-flavoured cheese sauce. Sprinkle with grated Parmesan cheese. Cook in the oven at 350°F (mark 4) for about 1 hour, or until cooked through.
To serve: Garnish with more Parmesan cheese if you wish, and accompany by French bread and green and tomato salads.

Macaroni cheese

12-oz pkt of 'quick' macaroni
4 oz mushrooms
a little butter

For the cheese sauce
4 oz butter
4 oz plain flour
2 pt milk
½–1 level tsp made mustard
8 oz mature cheese, grated
seasoning
¼ pt cider

To complete from freezer
4 fresh tomatoes

To make: Prepare the cheese sauce. Melt the butter in a pan, add the flour, stir well, then remove from the heat. Gradually add the milk, stirring all the time. Return the pan to the heat and bring to the boil, stirring continuously until the sauce thickens. Cook for a further 2 minutes, remove from the heat and cool slightly. Stir in the mustard, cheese, seasoning and cider.
Cook the macaroni in boiling salted water as directed on the packet and drain thoroughly. Sauté the mushrooms in the butter. Stir the macaroni into the cheese sauce.
To pack and freeze: Pour into 2 × 2-pt foil-lined dishes. Add the mushrooms, then cool quickly. Freeze rapidly until solid, then remove from the dish. Overwrap with foil, place in a freezer bag, label, seal and return to the freezer.
To use: Unwrap and return to the original ovenproof dish. Place in the oven at 350°F (mark 4), loosely covered with foil, for about 1½ hours. Remove the cover, arrange slices from 4 fresh firm tomatoes on top, dot with butter and return to the oven for 15–20 minutes.
To serve: Serve hot, garnished with parsley.

Liqueur cherry brûlée See page 29; make double quantity.

PÂTÉ AND POT LUCK FOR 8

Tuna pâté with shrimps

Beef and bean pot

*Onion and green pepper salad   Cottage loaves

Iced lemon pie

Starred dishes are recipe suggestions or accompaniments not suitable for freezing

Tuna pâté with shrimps

2 × 7-oz cans tuna, drained
8 oz full fat soft cheese
4 oz butter, melted
2 oz fresh white breadcrumbs
1 tbsp lemon juice
freshly ground black pepper
7-oz can shrimps, drained
4 oz butter, melted

To complete from freezer
lemon slices and parsley

To make: Place the tuna, soft cheese, 4 oz melted butter, breadcrumbs, lemon juice and seasoning in a bowl. Beat well with an electric beater until smooth and creamy. Chop the shrimps roughly and stir into the tuna mixture. Adjust the seasoning.
Spread the mixture in 8 individual ramekin dishes.

73

Smooth the tops and spoon over the remaining melted butter. Chill until set.

To pack and freeze: Wrap each ramekin in foil. Overwrap with freezer wrap, seal, label and freeze.

To use: Allow to thaw at room temperature for about 4 hours. Unwrap, garnish with lemon slices and parsley.

Beef and bean pot

3 lb lean minced beef
2 eggs
2 oz fresh white breadcrumbs
2 oz onion, skinned and finely chopped
salt and freshly ground black pepper
2 tbsps oil
3 × 15¼-oz cans red kidney beans
1-lb 13-oz can tomatoes
½ pt beef stock
5-oz can tomato paste
2 cloves garlic, skinned and crushed
3 level tbsps chili seasoning

To make: In a bowl combine the minced beef, eggs, breadcrumbs, onion, 1 level tsp salt and ground pepper to taste. Shape into about 80 small balls the size of a large walnut, using the palms of the hands.

Heat the oil, add a single layer of beef balls, cook over medium heat for 5 minutes to seal the surfaces. Repeat until all the meat is sealed. Drain well on absorbent kitchen paper.

In a large saucepan or flameproof casserole, combine the rest of the ingredients, including bean juices. Season with salt and pepper. Add the meat balls. Cover and cook over medium heat for 30 minutes. Uncover and continue to cook for a further 30 minutes until the juices have thickened. Stir occasionally to prevent burning on the base.

To pack and freeze: Cool rapidly, divide into two and freeze in preformed freezer bags or rigid containers. Overwrap in freezer bags, seal, label and return to the freezer.

To serve: Reheat from frozen in a covered casserole at 375°F (mark 5) for 1½–2 hours.

Cottage loaves

14 oz strong plain flour
1 level tsp salt
1 level tsp caster sugar
2 oz butter
⅓ pt milk
1 oz fresh yeast
25 mg ascorbic acid tablet, crushed
1 egg, beaten

Makes 12

To make: Sift the flour, salt and sugar into a basin. Rub in the butter. Warm the milk to tepid. Cream the yeast with a little milk, add the remaining milk and the ascorbic acid. Add the milk and beaten egg to the flour and stir well until the mixture leaves the sides of the bowl. Knead the dough for 10 minutes until smooth and elastic then divide into 12 equal pieces. Pinch off one-fifth from each. Shape both pieces into balls. Place a small ball on each large one. With the tip of the little finger press a hole in the small ball to attach the two, making small cottage loaf shapes. Place the loaves on a baking sheet; leave in a warm place until doubled in size. Bake at 375°F (mark 5) for 15–20 minutes. Brush with salted water and return to oven for a further 5 minutes. Cool on a wire rack.

To pack and freeze: Pack in freezer bags as soon as cold. Seal, label and freeze.

To use: Leave in packaging at room temperature for 1½ hours, then refresh quickly in oven.

Beef and bean pot

Iced lemon pie

8 oz digestive biscuits
4 oz butter
1 oz demerara sugar
3 eggs, separated
7 fl oz condensed milk
grated rind of 2 lemons
$\frac{1}{4}$ pt less 1 tbsp lemon juice
4 level tbsps caster sugar
$\frac{1}{2}$ gallon Soft Scoop vanilla ice cream

To make : Crush the biscuits finely. Melt the butter in a saucepan and add biscuit crumbs and demerara sugar. Press half the crumb mixture on to the base and sides of an 8-in pie plate, reserving about 3 level tbsps crumbs. Repeat using a similar dish. Beat the egg yolks until thick and creamy. Stir in the condensed milk, lemon rind and juice and stir until thick. Stiffly whisk the egg whites and whisk in the sugar until the mixture stands in peaks. Fold into the lemon mixture and divide between the crumb cases. Freeze until firm.

Scoop out small balls of ice cream and arrange on top of the pies. Decorate with the reserved crumb mixture.

To pack and freeze : Open freeze. When firm, pack in freezer bag. Seal, label and return to the freezer.

To serve : Remove from freezer a short time before serving.

Christmas with the Help of a Freezer

*Starred dishes are recipe suggestions or accompaniments
not suitable for freezing*

*Quantities or the number of servings are indicated
after the ingredient list of each recipe*

Chilled apple and apricot soup

2 lb apples, peeled, cored and chopped
8 oz dried apricots, chopped
2 pt light stock
$\frac{1}{4}$ pt dry white wine
$\frac{1}{4}$ pt soured cream

Serves 8

To make : Cook the apples and apricots in the stock until soft, then purée them in a blender. Add the wine and soured cream and chill.
To pack and freeze : Pour into a waxed carton or rigid polythene container leaving $\frac{1}{2}$-in headspace, seal, label and freeze.
To serve : Thaw in the refrigerator overnight and serve chilled.

Melon and ginger cocktail

1 honeydew melon ($2\frac{1}{2}$ lb)
2 oz caster sugar
$\frac{1}{4}$ level tsp ground ginger
juice and pared rind of 1 lemon

To complete from freezer
stem ginger

Serves 6

To make : Cut the melon in half, discard the seeds and, using a melon ball cutter, scoop out as many balls as practical. Place in containers suitable for freezing. In a saucepan heat the melon juice, made up to $\frac{1}{4}$ pt with water; add sugar, ginger, lemon juice and rind. Bring to the boil, stirring. Strain, and when cold add enough to cover the melon balls.
To pack and freeze : Cover the containers, seal, overwrap and label. Freeze until solid.
To serve : Partially thaw the melon by leaving it at room temperature, but serve it still chilled. Spoon a little chopped stem ginger over it.

Roast turkey with celery stuffing

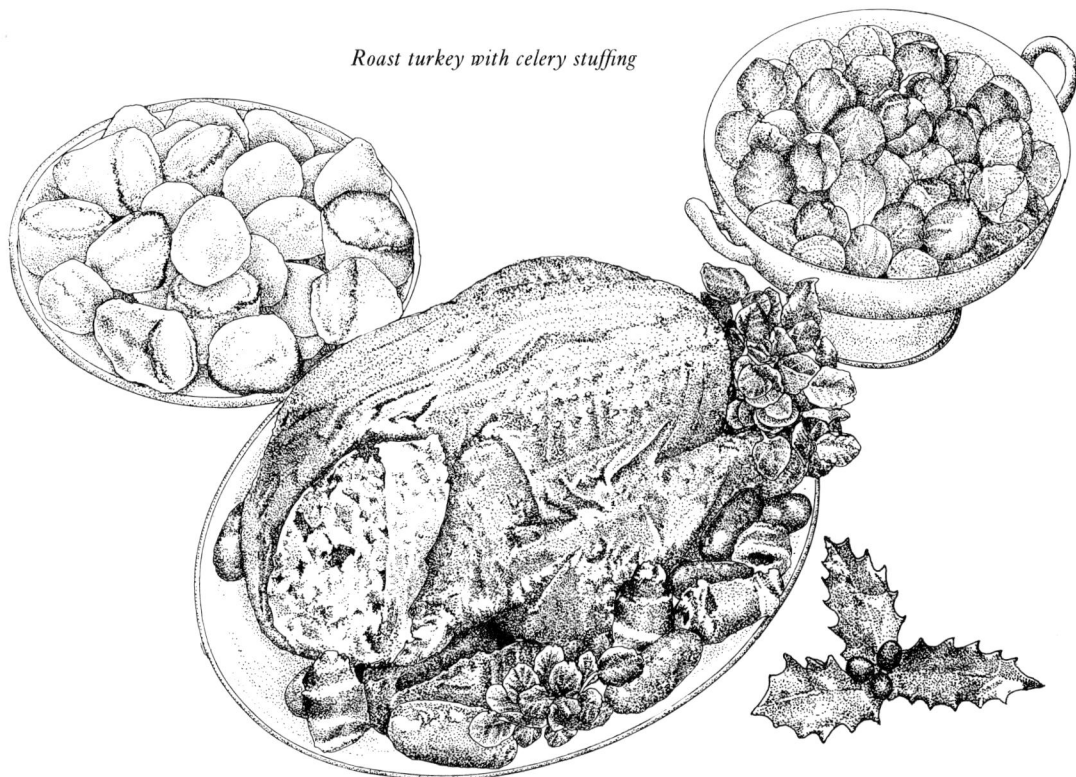

Thawing times
Allow about $1\frac{1}{2}$–2 days in a very cool place for up to 12 lb bird. Allow 2–3 days for larger birds.

Roast turkey with celery stuffing

Roasting times
Slow method (325°F, mark 3) with foil.
Quick method (450°F, mark 8) without foil.

WEIGHT (lb)	HOURS (slow)	HOURS (quick)
6–8	$3-3\frac{1}{2}$	$2\frac{1}{4}-2\frac{1}{2}$
8–10	$3\frac{1}{2}-3\frac{3}{4}$	$2\frac{1}{2}-2\frac{3}{4}$
10–12	$3\frac{3}{4}-4$	$2\frac{3}{4}$
12–14	$4-4\frac{1}{4}$	3
14–16	$4\frac{1}{4}-4\frac{1}{2}$	$3-3\frac{1}{4}$
16–18	$4\frac{1}{2}-4\frac{3}{4}$	$3\frac{1}{4}-3\frac{1}{2}$

How much for how many?
One 13-lb bird will serve 13–15 people.
A 16–20-lb bird will serve 20–30 people.
A frozen turkey of 9 lb dressed weight is equivalent to a fresh 12-lb turkey undressed weight.

Celery stuffing

4 oz onion, skinned and finely chopped
4 oz celery (2–3 sticks) finely diced
2 oz butter
4 oz cooking apple, peeled, cored and finely
    diced
6 oz breadcrumbs
1 small lemon, finely grated and squeezed
2 level tsps dried sage
salt and freshly ground black pepper
3 tbsps concentrated giblet stock

Sufficient for a 12–14 lb bird

To make: Sauté the onion and celery in the butter until transparent. Stir in the apple and sauté a little longer. Stir into the breadcrumbs with the finely grated rind, 1 tbsp lemon juice and the sage. Season well with salt and pepper. Bind all together with the giblet stock.
To pack and freeze: Place in a rigid container, seal, label and freeze. *Storage time:* 1 month.
To use: Allow to thaw overnight in the refrigerator and stuff the turkey in the morning.

Roast goose with mushroom and bacon stuffing

Thawing times
Allow about 24–36 hours in a very cool place.

Roasting times
Slow method (350°F, mark 4)
Quick method (400°F, mark 6)
Allow 25–30 minutes per lb by the slow method and 15 minutes per lb plus 15 minutes by the quick method.
Cover the bird with double thickness greaseproof paper during cooking but remove this 30 minutes before end of cooking time to allow the bird to brown.

How much for how many?
A 10-lb goose will serve 7–8 people.

Mushroom and bacon stuffing

6 oz fresh white breadcrumbs
1 tbsp chopped parsley
$\frac{1}{2}$ level tsp dried thyme
1 clove garlic, skinned and crushed
2 oz butter
4 oz onion, skinned and chopped
$\frac{1}{4}$ lb streaky bacon rashers, rinded and
    chopped
$\frac{1}{4}$ lb mushrooms, chopped
salt and pepper
egg yolk
stock, optional

Sufficient for about a 10-lb bird

To make: Put the breadcrumbs in a bowl and add the parsley, thyme and garlic. Melt the butter in a frying pan and sauté the onion until soft but not coloured. Stir in the bacon and cook a further 3–4 minutes. Stir in the mushrooms and mix in well. Add to the breadcrumbs. Season well with salt and freshly ground black pepper. Bind the ingredients together with an egg yolk and a little stock if necessary.
To pack and freeze: Pack in a rigid container, seal, label and freeze.
Storage time: 1 month.
To use: Allow to thaw overnight in the refrigerator and use to stuff the bird in the morning.

Roast duck with apricot and nut stuffing

Thawing times
Allow about 24–36 hours in a very cool place.

Roasting times
Allow 20 minutes per lb at 400°F (mark 6).

How much for how many?
Allow about 1 lb dressed weight per person.

Apricot and nut stuffing

6 oz fresh white breadcrumbs
4 oz dried apricots, finely chopped
2 oz salted peanuts, finely chopped
1 tbsp chopped parsley
2 oz butter
6 oz onion, skinned and finely chopped
1 small orange, grated and squeezed
$\frac{3}{4}$ level tsp curry powder
$\frac{1}{2}$ level tsp salt
freshly ground black pepper
1 small egg, beaten

*Sufficient for a 5–6 lb duck and a few stuffing
balls to garnish*

To make: Place the breadcrumbs in a bowl, add the
apricots, peanuts and parsley. Melt the butter in a
small saucepan and cook the onion and orange rind
gently, covered, until soft. Remove from the pan
and add to the breadcrumbs. Sprinkle the curry
powder into the pan and cook gently for 1 minute.
Pour over 3 tbsps orange juice; bubble gently for
30 seconds. Blend the curried juice into the
breadcrumbs. Season well with salt and pepper
and bind all together with beaten egg.
To pack and freeze: Pack in a rigid container, seal,
label and freeze.
Storage time: 1 month.
To use: Allow to thaw overnight in refrigerator and
use to stuff the bird in the morning.

Tarte noël

4 oz plain flour
pinch salt
2 oz butter, softened
2 oz caster sugar
2 egg yolks
6 level tbsps mincemeat
4 oz ground almonds
4 oz caster sugar
2 egg whites
almond essence
2 oz flaked almonds

To complete from freezer
4 tbsps apricot jam

Serves 6–8

To make: Sift together the flour and salt on to a
pastry board. Make a well in the centre and into it
put the butter, sugar and yolks. Using the
fingertips, pinch and work the pastry together
until well blended. Put the pastry in a cool place to
relax for 1 hour. Roll it out and use to line a $7\frac{1}{2}$-in
loose-bottomed French fluted flan ring and spread
mincemeat on the base.
Blend together the ground almonds, sugar, egg
whites and a few drops of essence. Pour this over
the mincemeat and cover with flaked almonds.
Bake the flan at 350°F (mark 4) for about 1 hour
until firm in the centre. Allow to cool in the tin.
To pack and freeze: Remove the flan carefully from
the tin, wrap in foil, overwrap in freezer wrap,
seal, label and freeze.
To serve: Unwrap and allow to thaw for 3–4 hours
at room temperature. Heat the jam in a saucepan
and brush over the almonds to glaze. Serve with
pouring cream.

Brandy butter

3 oz butter
3 oz caster or icing sugar
2–3 tbsps brandy

Serves 4–6

To make: Cream the butter until pale and soft.
Beat in the sugar gradually and add the brandy a
few drops at a time, taking care not to curdle the
mixture. The sauce should be pale and fluffy.

Rum butter

To make: Make this as brandy butter, but use soft
brown sugar, replace the brandy by 4 tbsps rum
and include the grated rind of $\frac{1}{2}$ a lemon and a
squeeze of lemon juice. Use as brandy butter.
To pack and freeze: Turn the brandy or rum butter
into the container in which it is to be served and
cover with a double layer of foil, or turn it into a foil
container. Overwrap with a freezer bag. Label
and seal.
To use: Allow to soften slightly for 1 hour at room
temperature. Use to accompany Christmas
pudding and mince pies.

Olde English creams

1-pt pkt lemon jelly
2 tbsps lemon juice
4 tbsps Olde English marmalade
$\frac{1}{4}$ pt single cream
$\frac{3}{4}$ pt double cream

To complete from freezer
blanched pistachio nuts

Serves 4–6

To make: Place the jelly cubes in a measure and make up to $\frac{1}{2}$ pt with boiling water. Stir to dissolve. Take out 5 tbsps of the jelly and mix with 2 tbsps cold water. Spoon this into a $1\frac{1}{2}$-pt jelly mould and leave to set. Add the lemon juice to the remainder of the jelly and chill until half set. Fold in the marmalade until well blended. Whisk the single and double cream together until the mixture will hold its shape. Fold evenly into the half-set jelly. Turn the mixture into the mould and leave to set.
To pack and freeze: Leave in the mould, cover with foil or freezer wrap and overwrap with a freezer bag. Seal and label. Freeze rapidly until solid.
To serve: Transfer to the refrigerator and thaw overnight in the mould. Before serving, unmould and decorate with blanched pistachio nuts.

Mince pies

12 oz shortcrust or flaky pastry
$\frac{3}{4}$–1 lb mincemeat
milk or egg to glaze
caster sugar

Makes 16 to 20

Using shortcrust pastry

To make: Roll out the short pastry to about $\frac{1}{8}$ in thickness. Cut into about 20 rounds with a 3-in fluted cutter, and 20 smaller rounds with a $2\frac{1}{4}$-in fluted cutter. Line $2\frac{1}{2}$-in patty tins with the larger

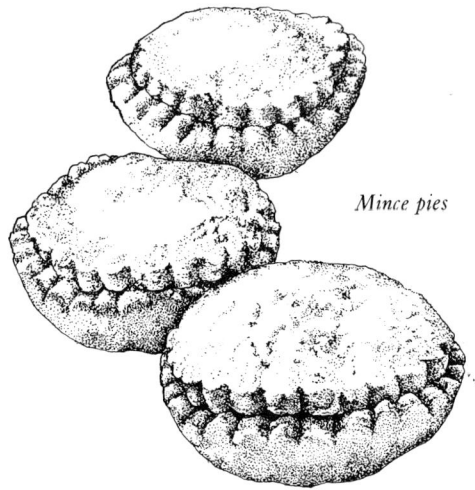

Mince pies

rounds and fill with mincemeat. Damp the edges of the small rounds and place in position.

Using flaky pastry

To make: Roll out the flaky pastry to $\frac{1}{8}$ in thickness. Stamp out 16 rounds with a $2\frac{1}{2}$-in plain cutter. Re-roll the scraps, cut another 16 rounds to use for the bases and place the bases on a baking sheet. Put a heaped tsp of mincemeat on each, damp the edges of the pastry, cover with the remaining pastry rounds and press the edges lightly together.
To pack and freeze: Leave the pies either in the patty tins or on baking sheets, unwrapped, until frozen firm. When quite firm, lift out from the tins or from the baking sheets; place in rigid containers with pieces of freezer paper between the layers and then overwrap, seal and label. Store in the freezer.
To cook: Take out the number of pies you need, return them to the original patty tins or baking sheet and cut a small slit in the top of each. Brush with a little milk and sprinkle with caster sugar. Bake in the oven at 425°F (mark 7) for shortcrust pastry, or at 450°F (mark 8) if using flaky pastry; cook for 20–30 minutes, or until golden brown and cooked through.
To serve: Sprinkle with more caster sugar, if you wish. Serve hot or cold.

Handy Standbys

Maître d'hôtel and garlic butters Ideal for topping steaks, chops and other grills such as fish. Cream 1 lb unsalted butter until smooth. Beat in the grated rind of 2 lemons, 4 tbsps lemon juice, 8 level tbsps chopped parsley and a little seasoning. Divide into 2, and add 4 skinned and crushed cloves of garlic to one portion. Chill both portions slightly in the freezer or refrigerator until of a manageable consistency. Form each piece into a long sausage shape between sheets of waxed or non-stick paper. Re-firm in the refrigerator or freezer, and cut into 2-oz portions. Wrap each in foil. (Alternatively, freeze the butter roll whole and cut off slices as required.) Make a neat parcel, seal and label. Pack the firm labelled parcels together in one freezer bag. Use straight from the freezer. (Store for up to 3 months for maître d'hôtel butter, 1 month for garlic.)

Mint butter Weigh out $1\frac{1}{2}$ oz mint leaves and $\frac{1}{2}$ oz parsley heads; wash and drain. Put in a pan with 4 tbsps water and cook uncovered for 5 minutes, by which time most of the water will have evaporated. Pass the mixture through a sieve. Cream 4 oz unsalted butter and work in the mint purée, a good squeeze of lemon juice and a pinch of salt. Form into a long sausage shape, and finish as above. (Store for up to 3 months.)

Butter balls take time to make when needed, but a few stored in the freezer – even those left over from a party – are handy for almost instant use. (Store for up to 3 months.)

Stock Particularly if you use a pressure cooker, the making of good-quality stock is a simple affair. Chicken bones, with skin and giblets, are well worth converting into a jellied stock; cooked meat bones, browned carrot, onion and celery, bacon rinds and mushroom stalks make a very useful concentrated stock. Or you can start from scratch with raw beef and veal bones. When the stock is ready, pour some of it into ice cube trays – useful when only a few tablespoons are needed – or use freezer bags set in rigid containers to give them shape. Freeze. (Store for up to 2 months as cubes, 3 months for larger amounts.)

Incidentally, the freezer is the ideal place to store a few odd trimmings and bones left when you have prepared or carved a bird, to wait till you have enough accumulated, or enough time, to turn them into stock.

Parsley and other fresh herbs may be frozen unblanched. Freeze clean parsley heads (pack the stalks separately – these are often required as a flavouring in recipes) and crumble whilst still frozen – to 'chop' the parsley. Alternatively, chop large amounts of crisp fresh parsley, or other herbs, divide between small containers (eg, ice-cube trays) and freeze until firm. Leave in the small containers or take from the ice-cube trays and pack carefully in rigid containers. Use straight from the freezer as an addition to stews, casseroles, sauces, etc. (Store for up to 6 months.)

Sauce cubes So convenient when only a small amount of sauce is needed at any one time. Make up a well-flavoured curry, tomato, espagnole or other sauce, using for preference a home made stock base rather than instant cubes. Cool, and pour into ice cube trays. Freeze unwrapped, divide into cubes and store in polythene bags. To use, select the number of cubes required and reheat slowly, preferably in an easy-clean pan. (Store for up to 2–3 months.)

Horseradish cream Whip $\frac{1}{4}$ pt double cream and fold into it 1 tbsp bottled creamed horseradish. Add a good squeeze of lemon juice and turn the mixture into a small serving dish or rigid foil container. Cover with double foil and overwrap with a freezer bag. Seal and label. To use, unwrap, but leave the top loosely covered, and allow to thaw for 6 hours in the refrigerator. Dust with paprika pepper before serving. (Store for up to 2 months.)

Mint sauce The leaves may be frozen whole, or chopped as for other herbs, and used to make up the sauce at the point of serving. However, for an almost instant sauce, you can chop enough mint to give $\frac{1}{2}$ lb, moisten with a little vinegar and add 1–2 oz caster sugar. Freeze in small amounts – ice-cube trays make ideal containers. When the mixture is frozen, pack the cubes in freezer bags or wrap individually in foil and then overwrap; seal and label. To use, add more vinegar, or one-third water and two-thirds vinegar, and leave to thaw. For a brighter colour, moisten before freezing with boiling water instead of cold vinegar, and proceed as above, adding undiluted vinegar to thaw. (Store for up to 6 months.)

Cranberry sauce Simmer 6 oz sugar and $\frac{1}{4}$ pt water in a pan until the sugar dissolves. Add $\frac{1}{2}$ lb fresh or thawed frozen cranberries and cook over a medium heat for about 10 minutes, until the fruit has broken up; cool. Pot, leaving a small headspace, cover with foil and freeze. (Store for up to 2–3 months.)

Breadcrumbs To prepare fresh bread-crumbs in quantity, especially with the aid of an electric blender, is a real time-saver. When frozen the crumbs remain separate, and the required amounts can be easily removed. Pack in freezer bags or sealed containers. (Store for up to 3 months.) There is no need to thaw crumbs to be used for stuffings, puddings and sauces, but for coating fried foods leave them at room temperature for 30 minutes.

Garlic and other savoury breads These go down so well when served with soups on a buffet occasion that it's worth stocking up before a party. Make 1-in cuts along French or Vienna loaves to within $\frac{1}{2}$ in of the base, open up and spread generously with creamed butter flavoured with garlic, cheese or herbs. Wrap the loaves tightly in foil, then freeze. Store for up to 1 week – the crust begins to shell off after this time. To thaw and refresh, place the frozen foil-wrapped loaf in the oven at 400°F (mark 6); a French stick takes about 30 minutes, and a Vienna loaf about 40 minutes.

Whipped cream rosettes or whirls These make an easy-to-handle decoration for all varieties of sweet dish.
Whip fresh double cream until just stiff enough to hold its shape. Pipe rosettes or shell shapes on to waxed or non-stick paper placed on a flat baking sheet. Freeze un-wrapped until firm; remove the shapes from the paper and pack carefully into rigid containers, with waxed or freezer paper between the layers; seal, label and return them to the freezer. (Store for up to 9 months.) To use, place in position on the dish to be decorated and allow ample time to thaw – preferably in the refrigerator for 1–2 hours.

Meringue shells Line a baking sheet with a sheet of greaseproof or non-stick paper. Whisk 2 egg whites until very stiff, add 2 oz granulated sugar and whisk again until the mixture regains its former stiffness. Lastly fold in 2 oz caster sugar very lightly, using a metal spoon. Put into a forcing bag fitted with a large fluted star nozzle. Pipe 'shell' shapes on to the lined baking sheet and bake in a very cool oven, at 250°F (mark $\frac{1}{4}$), until the meringues are firm and crisp, but still white. (If they begin to brown, prop the oven door open a little.) Allow to cool before packaging in rigid plastic containers for freezing. Can be used straight from the freezer for decorating desserts. (Store for up to 3 months.)

The following two recipes are not suitable for freezing but are used with menu suggestions in the book:

Mayonnaise

1 egg yolk
½ level tsp dry mustard
½ level tsp salt
¼ level tsp pepper
¼ level tsp sugar
approx ¼ pt salad oil
1 tbsp white vinegar

Put the egg yolk into a basin with the mustard, salt, pepper and sugar. Mix thoroughly, then add the oil drop by drop, stirring briskly with a wooden spoon or whisk, until the sauce is thick and smooth. When all the oil has been added, add the vinegar gradually – mix thoroughly.

French dressing

8 tbsps oil
2 tbsps wine or garlic vinegar
½ level tsp dry mustard
2 level tsps caster sugar
freshly ground black pepper
salt or garlic salt to taste

Measure all ingredients into a screw-top jar. Shake up just before using.

Béchamel sauce

4 pt milk
1 large carrot, peeled and roughly cut
1 medium-size onion, skinned and studded with 8 cloves
1 large sprig of parsley, including the stem
bouquet garni
1 level tsp salt
½ level tsp freshly ground black pepper
8 oz butter
8 oz plain flour

Makes 8 × ½-pt portions

To make: Heat the milk with the carrot, clove-studded onion, parsley, bouquet garni, salt and pepper. Bring almost to the boil, remove from the heat and leave to infuse for ½ hour. In a large pan melt the butter, but don't overheat it. Remove from the heat, add the flour and blend well together. Cook the roux for 3 minutes, remove from the heat and gradually add the warm strained milk. Cool quickly.

To pack and freeze: Divide into ½-pt portions and pack in preformed freezer bags; seal, label and freeze.

To use: Immerse the freezer bag(s) in hot water just long enough to loosen the contents. Slip them into an easy-clean pan and, over a low heat, stir until the sauce has softened. Bring to the boil and simmer for about 3 minutes, beating well to give a good gloss. Adjust the general seasoning and add additional flavourings as desired – for instance, capers, sherry, cheese, tomato paste, anchovy.

Tomato sauce

8 oz green streaky bacon, rinded and chopped
1 small onion, skinned and chopped
1 clove garlic, skinned and crushed
2 oz plain flour
2¼-oz can tomato paste
3 lb fresh tomatoes, skinned and seeded
3 pt chicken stock (fresh or cube)
bouquet garni
2 level tsps salt
freshly ground black pepper

Makes 6 × ½-pt portions

To make: Fry the bacon in a large saucepan until the fat begins to run. Add the onion, garlic and flour, cook for 5 minutes longer; blend in the tomato paste, tomatoes and stock and add the bouquet garni and seasonings. Stir well together, bring to the boil, then simmer for about 15 minutes. Cool quickly.

To pack and freeze: Pack in ½-pt portions in preformed freezer bags, seal, label and freeze.

Barbecue sauce

1 pt frozen bulk tomato sauce *(see previous recipe)*
2 level tbsps cornflour
2 tbsps water
1 tbsp vinegar
1 oz soft brown sugar
a dash of Worcestershire sauce

Makes 1 pack of ¾-pt and 1 of ½-pt

To make: Thaw the tomato sauce in a saucepan over a low heat. Blend the cornflour with the water to a smooth paste, add the vinegar, then stir into the warm tomato sauce. Add the sugar and Worcestershire sauce to taste. Bring to the boil and boil for a few minutes, stirring. Cool quickly.

To pack and freeze: Pack in preformed freezer bags in 1 × ¾-pt and 1 × ½-pt portion, seal, label and freeze.

Bolognese sauce

8 oz streaky bacon, rinded and chopped
2 oz lard
5 medium-size onions, skinned and finely chopped
3 lb best lean minced beef
2 × 5-oz cans of tomato paste
2 beef stock cubes
½ pt water
¼ pt red wine
salt and pepper

To complete sauce from freezer
1 clove garlic, skinned and crushed

To make: Fry the bacon in the lard, add the onion and cook for a further 5 minutes. Stir in the beef and cook until the meat begins to brown. Drain off the fat and reduce the heat. Stir in the tomato paste, crumbled stock cubes, water and wine. Add a little seasoning, cover and simmer for ¾–1 hour, stirring occasionally, until the meat is tender. Cool quickly. Use at once for lasagne or other pasta dishes, or freeze as follows:

To pack and freeze: Divide between three rigid plastic containers or lined preformers. Freeze rapidly until solid. Overwrap with foil or a freezer bag, seal and label.

To thaw: Reheat gently in a saucepan, breaking the mixture up with a wooden spoon as it thaws. Add 1 clove of garlic, skinned and crushed, during the reheating.

To serve: Serve with freshly cooked pasta such as spaghetti or pasta shells, or for lasagne.

Freezer Techniques

However keen to freeze you are, please bear with us while we give a brief description of how freezing works. It's a very simple process, and we don't want to make it sound complicated by going into details, but unless you grasp the principles, it's possible to make some equally simple mistakes. These may not make the food inedible, but they probably will affect its taste, texture and – most important of all if you're feeding a growing family – its food value.

Most foods are largely made up of water, and even something as 'solid' as lean meat contains about 70% of it. All freezing does is convert this water to ice crystals. *Quick* freezing makes tiny ice crystals, retained within the cell structure, so that on thawing, the structure is undamaged and the food value unchanged. *Slow* freezing results in large ice crystals, which damage the cell structure and cause loss of nutrients. As this damage is irreversible and slow-frozen food shows loss of texture, colour and flavour when thawed, you'll see why it is vital to follow the manufacturer's freezing instructions implicitly.

The first essential is never to freeze more than one-tenth of your freezer's capacity in any 24 hours. If, for instance, you have a 120 lb freezer (approximately 6 cubic feet), you should only ever freeze about 12 lb of food at a time. It's *possible* to freeze more, of course, but because the addition of the unfrozen food pushes up freezer-temperature, the results are going to be *slow-frozen* – the very thing you want to avoid.

The second essential is to freeze the food in precisely the way the manufacturer tells you to. Forget what you've seen other people do, because what's right for their freezer could be wrong for yours. Some freezers have a special rapid freezing compartment, and once the food's frozen, it has to be transferred to the rest of the freezer; with others it's necessary to freeze the food against the base or sides, or on whichever shelf carries the evaporator coils. Freezers need turning down to the coldest setting a few hours before you do any freezing. Beyond checking the temperature setting (it's worth investing in a small thermometer), there's nothing more to it – so now stop worrying and start freezing.

Wrapping everything up
As we have just pointed out, with freezing you have

Freezer packaging materials

85

Open freezing before packing

to do what you're told. There isn't much scope for bending the rules, and this applies particularly to packaging. Freezing converts water to ice within the cell structure of the food, and this is a state that *must* be maintained. If careless packaging allows further moisture to be withdrawn from other foods, or from the interior of the freezer, then frosting will appear on the exterior of the food. If this is allowed to continue, moisture will eventually be withdrawn from the food itself, causing dehydration and loss of nutrients. At its worst, the food could become so dried out that there wouldn't be any liquid left for it to re-absorb at the thawing stage. We're not suggesting you're going to fling the food in the freezer unwrapped (if you do, it will come out shrivelled and a fraction of its former weight), but it is tempting to get a bit slapdash with the sealing – and this is enough to do plenty of damage. With meat and poultry, for instance, exposed areas develop 'freezer burn': tissues go tough and spongy – so tough and spongy, in fact, that no amount of clever cooking can put it right. With any kind of food, strong smells can travel from one faulty package to another, so that a delicate lemon ice cream may come out reeking of curried beef. The longer you intend to store the food, the more important it is that you should package it properly. For short-term storage – 3–4 weeks – you might get away with cutting the corners, but the effects will show themselves as time passes. Now you know why proper packaging is so important, here's how to do it.

Package solids tightly, so that you expel as much air as possible. This is easy if you are wrapping something in aluminium foil, which fits where it touches, but it's not so simple if you are filling a rigid container, and run out of food half-way up. You can't squeeze the surplus air out, but you can fill up the vacant space with some crumpled non-stick, waxed or freezer paper. It's even more difficult to expel air from a freezer bag, especially when the contents are awkwardly shaped or easily broken. An effective and simple method is to lower the full bag into a bowl of water, so that the water-pressure can force the air up and out (remember to dry the bag before freezing, so you don't find yourself having to chip it out when you need it). Alternatively, use a drinking straw and suck the surplus air out.

Liquids are a different proposition, because instead of squeezing air out, it is essential to leave at least $\frac{1}{2}$ in of 'headspace' – up to 1 in for 1 pt. This is because water expands one-tenth on freezing – something you won't need telling if you had burst pipes last winter. Unless you leave room for expansion, soups, sauces or fruits packed in syrup will push off their lids; worse still, if you've used a screw-top jar (difficult to remove food from, and unless the glass is specially toughened, too fragile anyway), food could shatter its way out. Bear this in mind if you're ever tempted to chill a bottle of

Preforming using freezer bags

86

wine in an emergency: if you leave it too long, you will find it splattered around the freezer. If you *do* use a glass container, make sure it has straight sides – which make it easier to get the contents out again – and leave 1–2 in headspace.

Solids-plus-liquids, like stews and casseroles, or fruit in syrup, need to have a layer of liquid on the top, with no chunks of food sticking out. Remember to leave $\frac{1}{2}$ in headspace.

There are masses of packaging materials on the market, but many do the same job. You'll only need a few, so it shouldn't be too expensive, especially as many can be used again and again.

Heavy-duty (150-gauge) freezer bags can be washed, dried and used ad infinitum – though do check them thoroughly in case they've become punctured during use; ordinary polythene bags can be used for overwrapping. When using freezer bags for liquids, it's a good idea to fit the bag into a regular-shaped container before filling and freezing it, so you can slip it out when it is solid, neatly-shaped and ready for stacking. This is known as 'preforming', and any clean straight-sided container, such as an empty sugar carton or a plastic box (don't use glass), will do the job efficiently.

Just as some people swear by freezer bags, others find **aluminium foil** the best packaging. Ordinary thickness kitchen foil can be used double and wrapped fairly generously, but the heavy-duty quality sold specifically for use in freezers may be used singly. Any foil-wrapped packages are best overwrapped in thin polythene bags or freezer wrap, as foil punctures relatively easily. Heavy-duty foil is tough enough for re-use (soak it well, brush the food from the creases, and dry thoroughly), but you would be well advised to keep it for short term use only, once it is 'second-hand'. Foil is ideal for moulding round awkward shapes like joints of meat, fish and poultry, but add a freezer bag for extra security, if there is any danger of sharp bones piercing through. It is a good idea, too, to overwrap items intended for a longer-term storage.

Foil is also ideal if you want to end up with a neatly-shaped casserole instead of a space-wasting lumpy bundle. Of course, you *can* freeze food in its casserole dish (remembering that china and pottery slow the rate of freezing, and that the pack will need a lot of space), but if you want to keep the dish in use, line it with foil first, leaving a good margin for wrapping over; spoon in the contents and, once they've frozen solid, slip out the package and fold the foil over, then overwrap with

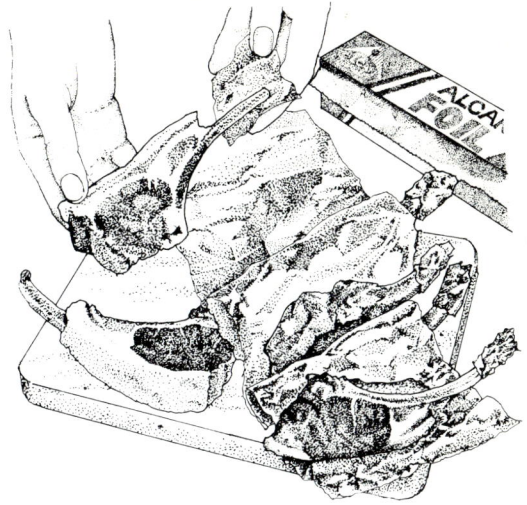

Protecting protruding bones before wrapping

freezer wrap before freezing. Slip the unwrapped package back into the original casserole dish for reheating. (If the foil tends to stick, dip the package for just a minute or two into warm water.) One proviso: unless the casserole dish is straight-sided and lipless, the contents may not 'slip out'. An alternative method is to cook the food in the casserole without foil, freeze it, then dip the casserole in warm water just long enough to loosen the contents, which are then removed and wrapped in the usual way.

Although foil can be re-used for freezing, in many cases (poultry, for instance), it makes sense

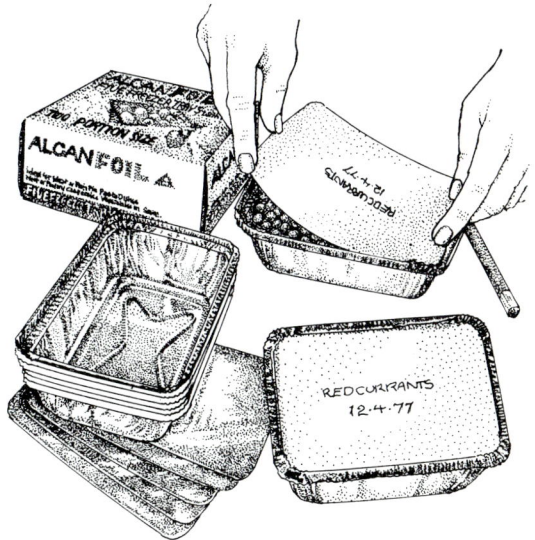

Foil containers with airtight and moistureproof lids

87

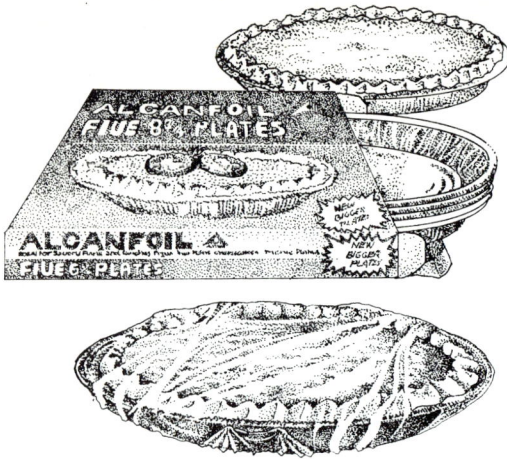

Overwrapping with freezer wrap

to loosen it slightly and use it for cooking the frozen food. This freezer-to-oven use explains why rigid aluminium foil dishes are so popular. You can cook, freeze and reheat in the same container, wash it up (soaking well, and brushing any food out from the angles), and start all over again for short-term usage. Some come complete with cardboard lids which press on firmly and are air and moisture proof. Before freezing, the dishes can be overwrapped with a double layer of foil, or placed in a freezer bag as an added protection.

Heavy duty polythene cling film of the quality sold for use in the freezer, is just as useful as foil for wrapping odd shapes; but of course it cannot be left on for cooking.

Freezer paper or waxed paper is useful for interleaving between individual portions in a bulk pack, as is foil or freezer wrap.

Waxed containers are less essential than freezer bags and foil – but they are handy to have around. If treated gently, it is possible to use them several times. Tub shapes are usually recommended for soups and sauces, but they waste freezer space, which is why some people stick to the squares and oblongs. A few words of warning: never fill them with anything hot, because the wax will melt and leave the container porous. Be cautious about using empty waxed cream containers instead of purpose made cartons. These are only very lightly waxed, and frozen food can start deteriorating inside them after a week or two. If box lids don't have an adequate seal, put a turn of special freezer tape round the join.

You'll have no such worries with **plastic containers**. These are almost indestructible and because they have snap-on lids, they're completely airtight. (When you reach the Sealing section below, you'll realise just how much of an advantage this is.) Rigidity means you can safely store fragile foods like vol-au-vents, cream-filled meringues and gâteaux in them – and though they may seem expensive at the time of buying, they'll more than earn their keep over the years.

Sealing
All kinds of parcels (whether they're wrapped in foil, freezer wrap, freezer paper or waxed paper) will have added protection by sealing with freezer

Clearly labelled packages

tape. This has a special adhesive that will stay stuck at freezer temperatures, but be generous with it. Use it for your waxed containers, too, unless they have screw-top, air-tight lids.

Labelling

Last but not least comes labelling – don't ignore it unless you're prepared to serve up apple pie instead of that chicken and ham you had in mind. It's easy to think you'll remember what's what, especially when you can still see the contents clearly through freezer wrap, but this soon frosts over and leaves you with a guessing game. Label your freezer bags, giving the date of freezing, the contents, the number of portions and – if you're feeling super-efficient – the reheating time. If your bags do not already carry a specially printed label, you will have to use a tie-on or stick-on waterproof label, securing it firmly. Write on tubs and cartons with a waterproof felt pen, or a chinagraph pencil. And to save yourself groping around in a chest-type freezer with icy hands, group foods-of-a-kind together in different coloured string bags.

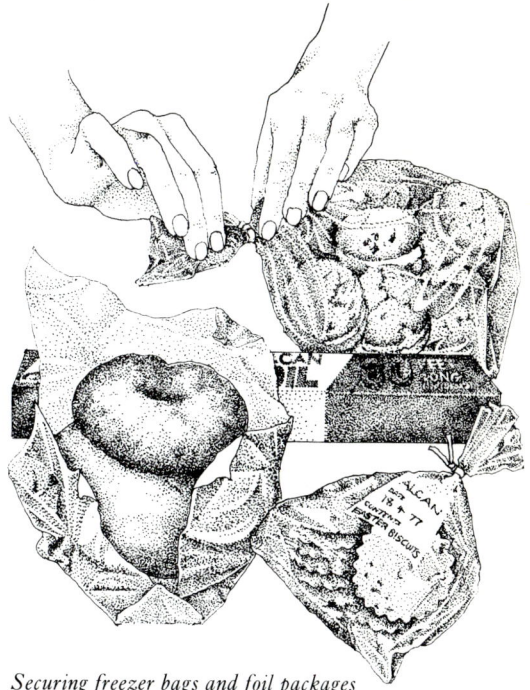

Securing freezer bags and foil packages

Guide to Catering Quantities

	Single portion	*24–26 portions*
Soup: cream, clear or iced	$\frac{1}{3}$ pt	1 gallon
Meat, with bone	5 oz	7–8 lb
boneless	3–4 oz	5–6$\frac{1}{2}$ lb
Poultry:		
turkey	3–4 oz (boneless)	16 lb (dressed)
chicken	1 joint (5–8 oz)	6 × 2$\frac{1}{2}$–3 lb birds (dressed)
Salad vegetables:		
lettuce	$\frac{1}{4}$	3–4
cucumber	1 in	2 cucumbers
tomatoes	1–2	3 lb
white cabbage	1 oz	1$\frac{1}{2}$ lb
boiled potatoes	2 oz	3 lb
Rice or pasta	1$\frac{1}{2}$ oz (uncooked)	2 lb
Cheese		
(for biscuits)	1–1$\frac{1}{2}$ oz	1$\frac{1}{2}$–2 lb cheese 1 lb butter 2 lb biscuits

	Makes 12 portions	*Makes 20 portions*
Mayonnaise	1 pt	1$\frac{1}{2}$–1$\frac{3}{4}$ pt
French dressing	$\frac{1}{2}$ pt	$\frac{3}{4}$–1 pt

Drinks

Drinking habits vary vastly. If you don't know the tastes and capacities of everyone coming to your party, a rough guide is:

Buffet parties

Allow for each person, 1 to 2 shorts and 3 to 6 longer drinks plus coffee.

Drop-in-for-drinks

Reckon on 3 to 5 short drinks each and 4 to 6 small savouries.

Drinks by the bottle

Sherry and port and straight vermouths give roughly 12–16 glasses per bottle. In single nips for cocktails, vermouths and spirits give just over 30 a bottle. Reckon 16–20 drinks of spirit from a bottle when serving them with soda, tonic or other minerals. A split bottle of soda or tonic gives 2–3 drinks. For table wines, reckon 5–6 glasses a bottle.

A 1-pt can of tomato juice will give 4–6 drinks. Dilute a bottle of fruit cordial with 7 pt water for 20–25 drinks.

	Single servings	24–26 servings	Notes
Coffee ground, hot	$\frac{1}{3}$ pt	9–10 oz coffee 6 pt water 3 pt milk 1 lb sugar	If you make the coffee in advance, strain it after infusion. Reheat without boiling. Serve sugar separately
ground, iced	$\frac{1}{3}$ pt	12 oz coffee 6 pt water 3 pt milk sugar to taste	Make coffee (half sweetened, half not), strain and chill. Mix with chilled milk. Serve in glasses
instant, hot	$\frac{1}{3}$ pt	2–3 oz coffee 6 pt water 2 pt milk 1 lb sugar	Make coffee in jugs as required. Serve sugar separately
instant, iced	$\frac{1}{3}$ pt	3 oz coffee 2 pt water 6 pt milk sugar to taste	Make black coffee (half sweetened, half not) and chill. Mix with chilled creamy milk. Serve in glasses
Tea Indian, hot	$\frac{1}{3}$ pt	2 oz tea 8 pt water $1\frac{1}{2}$ pt milk 1 lb sugar	It is better to make tea in several pots rather than one outsize one
Indian, iced	$\frac{1}{3}$ pt	3 oz tea 7 pt water 2 pt milk sugar to taste	Strain tea immediately it has infused. Sweeten half of it. Chill. Serve in glasses with chilled creamy milk
China	$\frac{1}{3}$ pt	2 oz tea 9 pt water lemons (27 thin slices) 1 lb sugar	Infuse China tea for 2 or 3 minutes only or it loses much of its fragrance. Put a lemon slice in each cup before pouring. Serve sugar separately

Freezer Storage Times

Meat, raw (leave unstuffed)
Beef: 8 months
Lamb: 6 months
Veal: 6 months
Pork: 6 months
Freshly minced meat: 3 months
Offal: 3 months
Cured and smoked meats: 1–2 months
Sausages: 3 months

Meat, cooked dishes
Casseroles, stews, curries, etc: 2 months

Meat, roast
2–4 weeks

Meat loaves, pâtés
1 month

Poultry and game (leave unstuffed; hang game
   before freezing)
Chicken: 12 months
Duck: 4–6 months
Goose: 4–6 months
Turkey: 6 months
Giblets: 3 months
Game birds: 6 months
Venison: 12 months

Fish, uncooked
Whole and fish steaks must be frozen really fresh
Salmon: 4 months
White fish: 6 months

Fish, cooked
Pies, fish cakes, croquettes, kedgeree, mousse,
   paella: 2 months

Sauces, soups, stocks
2–3 months; if highly seasoned 2 weeks

Pizza
Unbaked: up to 3 months
Baked: up to 2 months

Pastry, uncooked
Shortcrust: 3 months
Flaky and puff: 3–4 months

Pastry, cooked
Pastry cases: 6 months

Pastry pies, uncooked
Double crust: 3 months
Top crust: 3 months
Biscuit pie crust: 2 months

Pastry pies, cooked
Meat pies: 3–4 months
Fruit pies: 6 months

Pastries, Danish
Unbaked, in bulk: 6 weeks
Baked: 4 weeks

Pancakes
Unfilled: 2 months
Filled: 1–2 months

Sponge puddings
Uncooked mixture: 3 months
Cooked mixture: 1 month

Mousses, creams, etc
2–3 months

Ice cream
Bought: 1 month
Home made: 3 months

Cream
Up to 12 months, but ideally about 4 months

Cakes, uncooked, mixtures
Rich creamed mixtures only: 2 months

Cakes, cooked
Including sponge flans, Swiss rolls and layer
   cakes: 6 months (Frosted cakes lose quality
   after 2 months; since aging improves fruit
   cakes, they may be kept longer)

Scones and teabreads, cooked
6 months

Croissants
Unbaked, in bulk: 6 weeks
Baked: 4 weeks

Biscuits
Baked and unbaked: 6 months

Bread and rolls
Freshly baked: 4 weeks
Bought part-baked: 4 months

Sandwiches
1–2 months
Toasted: up to 2 months

Marmalade
Cooked pulp, without sugar: 6 months; useful if

it's not convenient to make marmalade when
Seville oranges in season

Butter
Unsalted: 6 months
Salted: 3 months

Eggs
Separated: 8–10 months

Herbs
Up to 6 months

Commercially frozen foods
Up to 3 months as a rule

Note: *The times quoted by the manufacturers are often less than those given for home-frozen foods, because of the handling in distribution, before the foods can reach your own freezer.*

Measuring in Metric

Guide to metric cooking
Use these equivalents to make metric recipes that will fit your cake tins and bowls.

Present measurement	Approx. metric equivalent	Present measurement	Approx. metric equivalent
1 oz	25 g	9 oz	250 g
2 oz	50 g	10 oz	275 g
3 oz	75 g	11 oz	300 g
4 oz	100–125 g	12 oz	350 g
5 oz	150 g	13 oz	375 g
6 oz	175 g	14 oz	400 g
7 oz	200 g	15 oz	425 g
8 oz	225 g	16 oz	450 g

Present measurement	Approx. metric equivalent	Present measurement	Approx. metric equivalent
1 fl oz	25 ml	15 fl oz	400 ml
2 fl oz	50 ml	20 fl oz	600 ml
5 fl oz	150 ml	35 fl oz	1 litre
10 fl oz	300 ml		

Quick handy measures

Approximate equivalent shown in level 15-ml spoonfuls			Approximate equivalent shown in level 15-ml spoonfuls		
Almonds, ground	25 g	= $3\frac{1}{2}$	Custard powder	25 g	= $2\frac{3}{4}$
Breadcrumbs, fresh	25 g	= 7	Curry powder	25 g	= 4
Breadcrumbs, dried	25 g	= 3	Flour, unsifted	25 g	= 3
Butter, lard, etc.	25 g	= 2	Suet, shredded	25 g	= 3
Cheddar, grated	25 g	= 3	Sugar, caster/granulated	25 g	= 2
Chocolate, grated	25 g	= 4	Sugar, demerara	25 g	= 2
Coffee, instant	25 g	= $6\frac{1}{2}$	Sugar, icing	25 g	= 3
Cornflour	25 g	= $2\frac{3}{4}$	Syrup or honey	25 g	= 1

Index